Learn AI and Human-Robot Interaction from Asimov's *I, Robot* Stories:
Robotics Through Science Fiction Vol. 2

ROBIN R. MURPHY

Copyright © 2019 Robin R. Murphy

All rights reserved. Do not distribute without permission.

CONTENTS

Introduction 1

Chapter 1 How Non-verbal Communication Influences Expectations and Trust: "Robbie" 9

Chapter 2 Why the Three Laws of Robotics Do Not Work in Real Life: "Runaround" 21

Chapter 3 Mental Models and How They Impact Interaction and Trust: "Reason" 35

Chapter 4 Transparency, Visibility, and Attribution: "Catch That Rabbit" 47

Chapter 5 Full Moral Agency: "Liar!" 56

Chapter 6 How Robots Are Programmed: "Little Lost Robot" 66

Chapter 7 User Interfaces: "Escape!" 77

Chapter 8 The Uncanny Valley: "Evidence" 89

Chapter 9 Autonomy, Initiative, and Trust: "The Evitable Conflict" 97

Chapter 10 Conclusion 106

ACKNOWLEDGEMENTS

I would like to sincerely thank Lilas Dinh, Andrea Lawrence, Dilip Patel, Jeff Craighead, Joanne Pransky, Joseph Waddington, Linda Ravey, Matthew Long, Mohsen Aghashahi, and Ramviyas Nattanmai Parasuraman for their early support of this book. Their sponsorship made me even more excited about the project and let me visualize my audience. I am also grateful to have Jennifer Reinoehl of Paradoxical as a copy editor and Amazon publishing support guru. She caught a lot of awkward writing, grammatical atrocities, inconsistencies, and boring repetitions; any that remain are totally my fault. I also appreciate discussions with Erika Rogers, who chaired the 2001-2002 DARPA/NSF workshop and study on human-robot interaction that established the field of HRI. This book was pitched through Publishizer, a crowd sourcing literary agency, and although that did not attract the attention of a large publishing house, I do appreciate the advice and time spent by Lee Constantine.

Introduction

I.1 The Purpose of This Book

The world is becoming full of robots: warehouse robots, robot museum guides and receptionists, small drones buzzing overhead, self-driving cars, telepresence robots, and planetary rovers, just to name a few. The robots are becoming increasingly intelligent whether they are advanced vacuum cleaners or Sophia, the humanoid robot granted citizenship in Saudi Arabia. What can regular people who are trying to get through the day expect from these robots? Will robots gain sentience and take over the world or can they really be programmed with Asimov's Three Laws of Robotics as a safety net? Why are humanoid robots so creepy to Westerners? These questions and others are the subject of research and development in artificial intelligence (AI) and human-robot interaction (HRI).

Human-robot interaction is not a subset of artificial intelligence or robotics; it is a multi-disciplinary field that investigates how people and robots work together, either implicitly or explicitly. AI includes how people react subconsciously to robots, how they communicate with robots, and what they fear or find annoying about robots. As robots become more ubiquitous in everyday life, designing appropriate human-robot interactions is becoming critical to the commercial success of robots. Although robots have many more dimensions than a simple mobile app, HRI can be thought of in the same way as the user friendliness of a mobile app: No matter how useful a mobile app is, consumers will discard it if the user interface is too cumbersome.

Not only is a robot more complex in terms of programming but our relationship to robots is also more complex than a user interface for a mobile app. Consider the different ways that humans and robots may interact. They may physically interact with each other, such as a robot bringing parts to a person in a warehouse or a robot assisting a disabled person. They may cooperate cognitively to share portions of a task, such as the computers that assist drivers through antilock braking systems and fuel economy controls or NASA scientists collecting data with a planetary rover. Humans and robots may also interact socially or emotionally, such as a robot working with an autistic child to help him or her make eye contact or a telecommuting robot serving as a surrogate for a remote worker. As humans and robots interact, humans have to trust that the robot will perform as expected.

This book is an introduction to the field of human-robot interaction using the stories in Isaac Asimov's *I, Robot* anthology as imaginary case studies that capture the complexity of our relationships with robots. It is intended for two different audiences. One audience is the layperson who has not taken college level classes in the field, but is interested in learning about AI and the implications of robot interactions. The book strives to provide these laypeople with a look at how robots might be used and the challenges in making interactions with them safe and pleasant—without the hype seen in the popular press. The second audience is novice human-robot interaction practitioners, such as engineers with a poor background in psychology and communications, psychology or communication scientists with little AI training, or legal experts unfamiliar with robotics. It is difficult to learn a new field of science from textbooks or journal articles targeting upper level students who are majoring in the subject. Instead of a textbook approach, this book exposes novices to multi-disciplinary topics in HRI and reinforces why these topics are important using familiar stories and simple language. The book can be used as a supplemental text for a human-robotics interaction course.

Case studies are a fantastic way to explain complex systems, but there are relatively few examples to draw upon in this emerging area of technology. The real-world studies that exist rarely describe a complete system or application. On the other hand, Isaac Asimov famously created entertaining stories imagining the edge of the robotic revolution on which we live. In his stories, he further realized the conflicts of interest that might occur and created the Three Laws of Robotics, which the popular press and government officials currently use when discussing robots.

Asimov's *I, Rob*ot anthology serves as a *de facto* set of case studies. It consists of nine short stories written in the 1940s and 1950s about robots designed for a variety of missions, from specialized applications (e.g., mining, running a space station) to roles involving human-level intelligence and social skills (e.g., a child's caregiver, an assistant technician, an elected government official). Asimov was able to connect the stories under the frame story of Dr. Susan Calvin, the lead robopsychologist for U.S. Robots and Mechanical Men, Inc., giving an interview on the 50th anniversary of her retirement. The stories anticipate technological advances scientists are currently developing that are not yet commercially available. Similar to a real case study, each story concentrates on how the robots would be integrated into society, and, in particular, why a presumably logical robot, bounded by the Three Laws, did something unexpected. In the stories, an ambiguous command from a human frequently causes the unexpected action and human troubleshooters struggle to find the cause of a problem that is concealed by a lack of design transparency. Thus, these stories are excellent counter-examples of good human-robot interaction design.

Furthermore, Asimov created stories with a positive view of intelligent robotics: None of the robots are staging an uprising and weaponized robotics are not an issue since mankind has moved beyond war. The robots are always trying to do the right thing, and when they do things wrong, it is usually portrayed humorously. This positive view allows us to critically analyze each case. Although the stories are slightly dated and, like the TV series *Mad Men,* they reflect the sometimes shocking stereotypes and attitudes of the 1940s and 1950s, they are still very relevant. Many of those dreams of the past are the goals toward which modern scientists currently strive when designing the robots of today.

I.2 How to Use This Book

To use this book, you must have a copy of *I, Robot* or *The Complete Robot* (the latter book contains all the stories from *I, Robot* plus all the other robot stories Asimov wrote after 1950). Chapters 1-9 in this book are paired with the nine stories in *I, Robot*. Reprinting the stories within this text would not be practical for copyright, cost, and space considerations especially since they are readily available.

As you approach each paired chapter, begin by reading the "Before You Read" and "As You Read" sections in this book for the particular chapter. The "Before You Read" section introduces you to the key scientific concepts of interest in the story. "As You Read" alerts you to specific instances of human-robot interaction that you should reflect on as you read the Asimov story. It presents you with questions like, "Why did the robot do that?", "Do I expect robots to behave that way?", and "Can robots do that in real life?"

After you have prepared, you can read the Asimov story, thinking about the questions as you read. Once you have finished, return to the "After You Read" section in this book. This section expands upon the science behind the story and explains how that instance of HRI would actually be implemented with AI. That section also summarizes the HRI lessons found in that story and restates them as one or more principles.

Each chapter concludes with questions to encourage self-reflection or group discussion in a classroom setting. Once you finish going through the first nine chapters in this manner, the final chapter summarizes what was learned, reviews the 12 principles derived from the stories, and briefly discusses other aspects of HRI that Asimov did not cover in *I, Robot*. The book concludes with recommendations for further reading.

The chapters in this book are designed to be read in sequence. The material in each chapter builds upon what was learned in previous chapters. If you are a science fiction aficionado, you may have already read these stories; however, in order to concentrate on the instances of human-robot interaction that serve as teachable moments, it is important to reread them as you are working through the chapters of this book

I.3 Overview of Artificial Intelligence and HRI

Most of us have come in contact with robots either in real life or through media. However, the concept of artificial intelligence is more difficult to understand. Artificial intelligence has many definitions, most of which are hotly debated. With the current emphasis on machine learning, many definitions in the popular press include a robot's ability to learn new things. But learning is not essential for some intelligent robot applications. One example would be a robot vacuum cleaner or a mower that covers an area randomly in the same way a sheep eats. These machines are smart and rely on AI even though they never learn the layout of the house or yard. The easiest way to think of artificial intelligence is having a computer or robot perform something that we would consider smart.

If we are going to understand how robots interact with humans, we need to understand some of the fundamentals of AI just as we should already understand quite a bit about fellow humans. To begin to understand the types of AI used in robots, we must look at the hidden phases in a robot task, how software is organized, and how robots are tested. Chapters 3, 5, and 7 touch on these topics, especially the things robots do as a result of programming limitations when a novel situation occurs.

We also need to understand more than just the AI on the inside of the robot enabling it to perform intelligent actions, and delve into how AI enables robots to communicate, both verbally and non-verbally. Communication comes naturally to humans, but it reflects very sophisticated mental models of action, learning, control, and reaction. AI researchers who focus on understanding natural language currently do not know how to create these mental models in robots, except for very in specialized applications. Chapters 2 and 4 address these communication challenges.

Another equally important aspect of AI for HRI is defining and creating the intelligence needed for ethical behavior. Robots must always behave in a safe manner when they interact with humans. They also must have user-friendly interfaces. Both of these requirements encourage positive HRI. Chapters 6 and 8 discuss the importance of these aspects of AI and how they might be implemented.

Beyond simply programming artificial intelligence, there are two aspects of human-robot interaction of which we should be aware. Some robots trigger a negative, visceral reaction in Westerners; people from these cultures often have a subjective human reaction that a particular robot is inherently creepy. If robots are to be a part of our society, this reaction must be minimized or suppressed. The "Uncanny Valley" is a concept that describes this reaction and explains in general terms how to prevent it. This aspect of HRI is discussed in chapter 9.

The second aspect is that, along with the Uncanny Valley, interactions with robots must be guided by a certain level of trust. Trust is another subjective human trait. Physical features, including movement, imply the competence of a robot, which influences trust. But other aspects such as mathematical proofs that a robot will operate correctly or has a history of reliability serve as the character reference for a robot, and that also influences how well a human will trust it. Trust as an aspect of HRI is discussed in chapter 10.

From the descriptions above, it is clear that the science behind what humans want to experience during an interaction with a robot goes beyond the topics normally covered in artificial intelligence or robotics courses. HRI shares research methodologies with human-computer interaction (HCI), but unlike computers robots move in the real world. They are physically situated agents. HRI also leverages research in communication, psychology and cognitive science, and ethics.

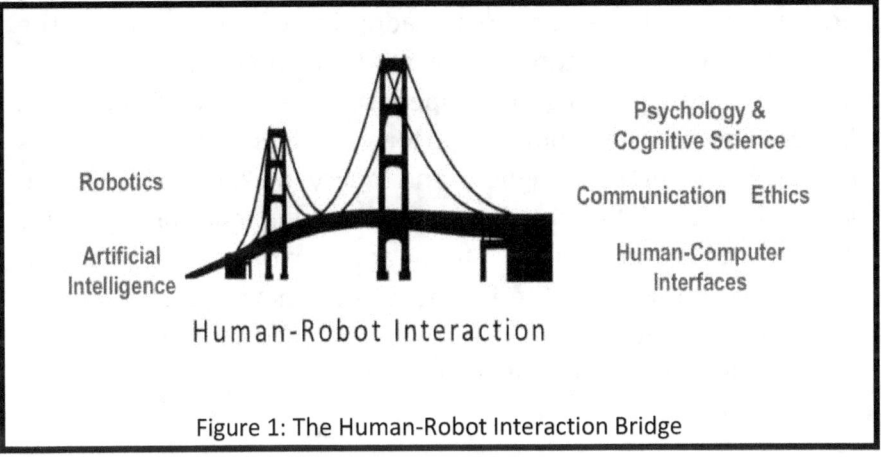

Figure 1: The Human-Robot Interaction Bridge

As illustrated in Figure 1, this requires human-robot interaction practitioners to create a bridge between multiple disciplines. Such bridges are usually difficult to construct because each discipline has its own vocabulary, research methodologies, and techniques. Even when someone is able to bridge the different disciplines, their expertise in human-robot interaction may not be viewed as essential. For example, a robot start-up company may not have an incentive to do more than create a minimum viable prototype in an attempt to get market share, falsely assuming that all the difficult "user stuff" is extra and can be added on later by a traditional programmer. Unfortunately for the start-up, a robot that violates known principles of how people think and react and violates the expectations of human ethics is subject to both rejection by consumers and product liability lawsuits. Therefore, the more designers (and regulators) understand HRI, the more safe and trustworthy robots will be produced.

I.4 Next Steps

Robots can have profoundly positive impacts on humankind if we pay attention to how we program the robot to support good human-robot interactions. Care in the design and regulatory process can facilitate commercial adoption of robots and, at the same time, prevent future disasters, such as predictable accidents due to drivers failing to supervise their self-driving cars. Understanding what humans subconsciously expect from robots can also aid future designers and policy makers to overcome irrational fears. Without explicitly designing and planning for human-robot interactions, trial and error could lead to devastating outcomes. The first step is learning the essentials of human-robot interaction.

Now, let's get started!

Chapter 1
How Non-verbal Communication Influences Expectations and Trust: "Robbie

"Robbie" is perhaps the most beloved robot story of all time. It is the story of a girl, Gloria, and her humanoid playmate, Robbie. Robbie is the perfect companion and protector. He cannot speak or even beep like the familiar Star Wars "droid" BB-8, but he communicates non-verbally in ways Gloria understands perfectly. Although the relationship between Gloria and Robbie is affectionate and his every action is clearly in Gloria's best interests overall, Gloria's mother does not trust the robot. These aspects of the story make "Robbie" a thought-provoking introduction into the human-robot interaction topics of communication and trust.

1.1 Before You Read "Robbie"

People often assume that having verbal dialog with another person is easy, forgetting that it takes more than 10 years of constant practice and coaching for a child to reach an acceptable level of verbal interaction. Because verbal interactions are normal for humans, from a baby's first cries to a dog's barks, we respond verbally and expect robots to also manage some form of it. A Star Wars robot without the "Droidspeak" language beeps would make for an entirely different movie.

The mechanism of verbal dialog where a person tells the robot what they want it to do or converses with it in a back and forth manner is called "natural language understanding" in artificial intelligence. The term "natural language" is used to distinguish human languages (English, Spanish, Russian, Mandarin, etc.) from a computer programming language like C++ or Python. Creating robots that understand natural language is considered one of the most difficult disciplines in AI.

Taking communication a step further and creating a robot that not only understands natural language but also interacts using it is even more complicated. This is because communicating through natural language is not the only way humans or animals communicate. The powerful combination of verbal and non-verbal forms of communication is often called "multi-modal communication." We smile, frown, make eye-contact, shift our gaze to objects when we talk about them, and speak in an angry or encouraging tone that even babies and dogs can understand. These forms of communication are non-verbal and typically serve to amplify what is being verbally communicated (through gesturing, pose, etc.) or express mental states (through facial expressions, tone of voice, sighing, etc.). Non-verbal communication is so important that it can be sufficient by itself. One example of this is a house-trained dog that communicates with its master when it wants to go outside.

Humans interpret non-verbal communication cues to determine whether or not another person or animal is trustworthy. Trust is influenced by observing whether the other agent is meeting expectations of appropriate behavior for the situation and acts predictably and reliably over time or not. In movies, we know the killer is creepy because their non-verbal cues do not match what he or she says. Consider in *Silence of the Lambs* when the FBI agent, Clarice, first meets Hannibal, a serial killer, in his jail cell. Hannibal stares intently at Clarice while he speaks with her, as if she is prey and he is the hunter. He winks slow and ominously. Although he is smiling and conversing with her in a polite manner, moving his head and eyes slightly to indicate when he is finished talking in order to signal it is her turn, the audience would not trust him even if the clear acrylic wall of the cell separating them and the rest of the jail setting did not contribute to the threatening feel of the moment. Subconsciously we have the same fearful reaction to a humanoid robot that does not blink, moves and speaks slowly and deliberately, and does not indicate that it knows whose turn it is to talk in the conversation.

The need for non-verbal communication puts a great burden on human-robot communication because people subconsciously expect a talking robot to give non-verbal cues that match what it is saying. Mobile robots are expected to use posture and poses that indicate they are aware of us. They should slow down and act non-threatening when they enter our personal space. Creating a robot that gives non-verbal cues to match its verbal dialog is hard because it involves autonomic expressions of internal state that are not performed by a conscious effort in humans (e.g., blushing when surprised), it involves coordination of these autonomic expressions with verbal dialog, and it involves reasoning about the content.

Consider a new employee who walks into a room with his boss as she is working behind her desk. He says, "Hello, are you busy?" If his boss were not busy, she might say something like "Come in," using a friendly tone and look up from her work to gaze at the man. On the other hand, if she says "Come in," but uses a neutral tone and does not look up, the man knows his boss will speak with him, but he should wait until she looks up from her work. The non-verbal cues reveal more about the internal state of the boss than the words.

Also notice that the non-verbal cues in the example above are synchronized with the verbal actions. The tone of voice (prosody) has to be coordinated for "Come in," as a phrase. The boss could not say "Come," in a friendly manner and then switch to a neutral or angry tone for the word "in." There also should not be any latency between saying "Come in" and looking up to make eye contact, otherwise, the combination of words and actions goes from being friendly to signaling annoyance. Non-verbal expressions of internal state and prosody are hard to program into robots because they have to deliberately reason about the effect they want to produce.

Continuing with this comparison between a human and robot, suppose the man resumes the conversation with "Pardon me, but where is Room 209?" The woman may respond in a compassionate or irritated voice and would say something along the lines of "You need to go back to the elevator and follow the signs," while pointing in the direction of the elevator. To duplicate this feat of multi-modal communication with a robot, it would have to recognize objects (the man), understand what the man needs, reason spatially (the path to 209), and understand the context of the situation on which the gesture should be based (which is so hard to program a robot to understand that it has its own area of study called deictic gesturing). It would not be beneficial to the man if the robot pointed directly to room 209 instead of toward the elevator.

Multi-modal communication impacts trust in a robot, especially one that is expected to function in social situations such as Robbie playing with Gloria. Trust in a robot is different than trust in an intelligent software package running on the Internet. Software generally offers an "undo" or "back" function and requires confirmation before executing important commands, such as spending money or signing agreements. A robot is a physical entity that performs autonomous actions that may be irreversible. For example, an early autonomous janitorial robot crashed through a wall in an office building, which undermined trust.

Trust is larger than consistency in multi-modal communication. For example, real robots include small drones, self-driving cars, automated warehouse robots, and other types that do not have a humanoid appearance. How do we develop trust in them? Scientists are not entirely sure how, but *"Robbie"* gives us some clues. One way to foster trust is to have robots certified safe by a higher authority. In this case, the company must provide a mathematical proof of correct operation and must be legally liable for the actions of their robots.

Another way that users develop trust is by direct observation of the robot's reliability and predictable behavior, in much the same way we develop trust in a new pet or a new employee. The robot, Robbie, is reliable and trustworthy in that he would not harm Gloria. However, we do see that Robbie is deceptive as a part of his play. For example, he slows down so Gloria can beat him to the tree. He also "looks" for Gloria even though he has identified her hiding spot. If anything, these internalized deceptions increase Gloria's trust in him as a friend. They also establish predictable behavior. Gloria knows she will have an advantage in any game. If instead of slowing, Robbie had increased its speed as he neared the tree, this would have destroyed some of Gloria's trust. Note the role of context in trust. It is currently very hard to program a robot to understand what the current context is and what is appropriate for that situation.

Another big push to increase trust in artificial intelligence is "explainable robotics." This branch of robotics enables the robot to explain what it is doing and why. The idea of explainable AI is very old: It started with the first practical expert systems in the 1970s. At that time, a rule-based program called MYCIN was making great strides in identifying the cause of a bacterial infection and recommending the correct antibiotic. It was like the computer Watson winning at Jeopardy. But doctors found bugs from time to time, as would be expected. Of course, the doctors wanted to update the rules of the program themselves as soon as they found them. But they could not find or understand what rules had led to MYCIN's conclusion nor could they predict the consequences of tweaking a rule or adding a new rule. Because of this, researchers wrote a new program called to make the rules and the interactions between those rules visible. The program was called TEIRESIAS after the character in the play Oedipus Rex who tries to explain to Oedipus that the oracle's prophesy that he will kill his father and marry his mother is correct. There is some black humor in the name TEIRESIAS because Oedipus did not listen to Teiresias, and there are indications that people will not listen to explanations generated by programs. When you include a feature in an artificial intelligence system that allows everyone to review the processes, it is called making the knowledge base visible or transparent.

Robbie does not appear to have the ability to explain its actions. However, Gloria's mother could watch Robbie's everyday interactions and see that they were predictable and reliable. Despite proofs, liability, and direct observation Gloria's mother does not see Robbie as trustworthy.

1.2 As You Read "Robbie"

As you read the story, watch for the different forms of non-verbal communication that both Robbie and Gloria use. Is the non-verbal communication signaling a general internal state that reveals how the agent is feeling independently of whether it is involved in a give-and-take conversation? Or is it part of a conversation between two or more agents, called dyads, where the non-verbal communication is synchronized with the conversants taking turns. Make a list as you read of the ways in which Robbie shows he is trustworthy. Think about the differences between trust in a computer program and trust in a robot. Also consider what it would take to make you believe a robot is trustworthy. At what point in the story would you trust Robbie?

1.3 After You Read "Robbie"

The Venn diagram in Figure 1 shows the current practice in robotics is to use traditional user interfaces with some systems providing natural forms of communication, mostly verbal. Traditional user interfaces are devices like joysticks or graphical user interfaces, such as the screen on a computer or mobile device. In human-to-human communication, humans use both verbal and non-verbal mechanisms to communicate. In human-to-animal communications, humans continue using both verbal and non-verbal forms, but animals use only non-verbal communication. Whereas Robbie used only non-verbal communication, modern robots and AI agents such as Siri and Alexa that incorporate natural forms of communication, typically use only verbal communication. There has been much research about adding non-verbal cues, but in general robots are not designed to solely exhibit non-verbal communication like Robbie. At one time, there was an industrial robot named Baxter, from Rethink Robotics, which would nod its head to confirm that the operator had correctly stored a new task, but it did not use verbal communication. Factory workers liked Baxter because it was made teaching situations more natural, but unfortunately the robot's physical motors had problems in accurately executing motions and Rethink Robotics went out of business.

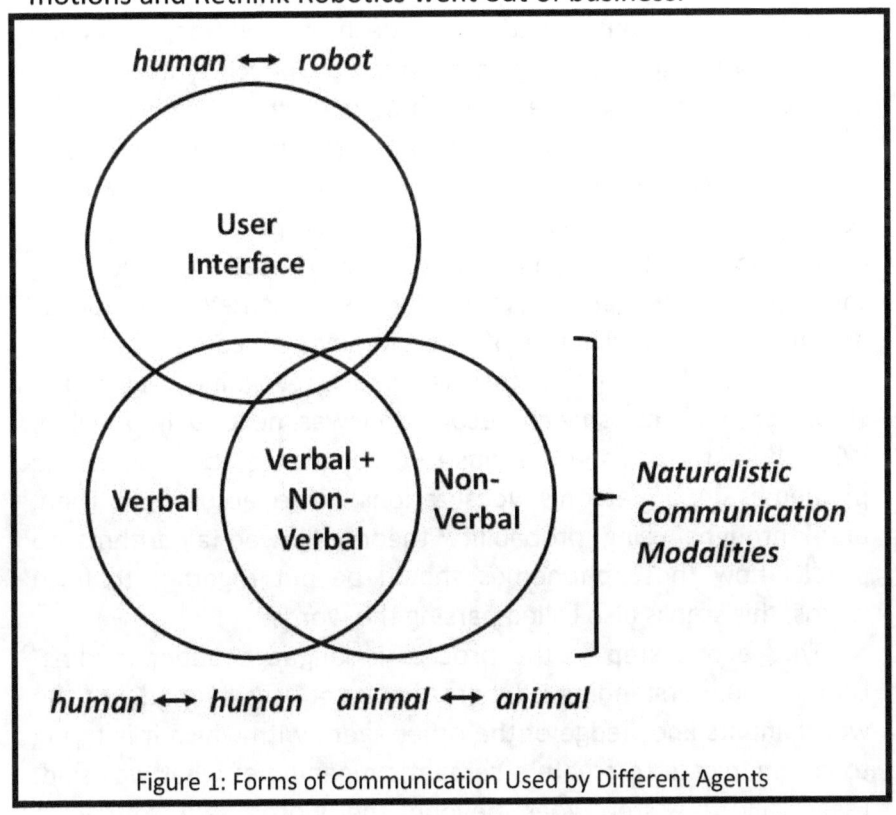

Figure 1: Forms of Communication Used by Different Agents

Surprisingly it is easier for a robot to be programmed with verbal communication and generate verbal responses than it is to create a robot that uses non-verbal communication. Speech generation is straightforward—the hard part is determining what to say. For example, once a robot understands that it is being asked directions to a particular place, it is easy to compute a path and then put the steps of that path into words. After all, any online map service does this for you. However, whereas programming the output for a simple request for directions is easy, programming a robot to listen to a human and convert the variety of different sounds and pronunciations a person makes into actual words, and then to pick out the important words so that the robot understands what the human is communicating is very difficult.

Researchers generally break natural language communication into three different functions: speech recognition, language understanding, and language generation or speech synthesis. This can also be thought of as the input (speech recognition), the information processing (language understanding), and the output (language generation or speech synthesis). Figure 2 shows the general process. You might notice that incorporating non-verbal communication cues is not directly referenced. Incorporating these cues, as well as understanding overall context of requests, are largely contained in the internal workings of "context."

In natural language processing, speech recognition is the first step. For decades, speech recognition was next to impossible. With breakthroughs in acoustics, computers can hear the phonemes in a person's vocalizations more accurately. Then, breakthroughs using probability theory allowed algorithms to predict how those phonemes should be put together to form words; this step is also called parsing the words.

The second step in the process is language understanding. Language understanding is where the agent uses its model of the world and its knowledge of the other agent with which it is trying to communicate to create a semantic meaning of what was said. Essentially this step goes beyond the words said to try to understand the real, underlying intent. Understanding meaning is called semantics and is very, very difficult for robots to perform.

Once the words are known, algorithms use knowledge about syntax to put together sentences.

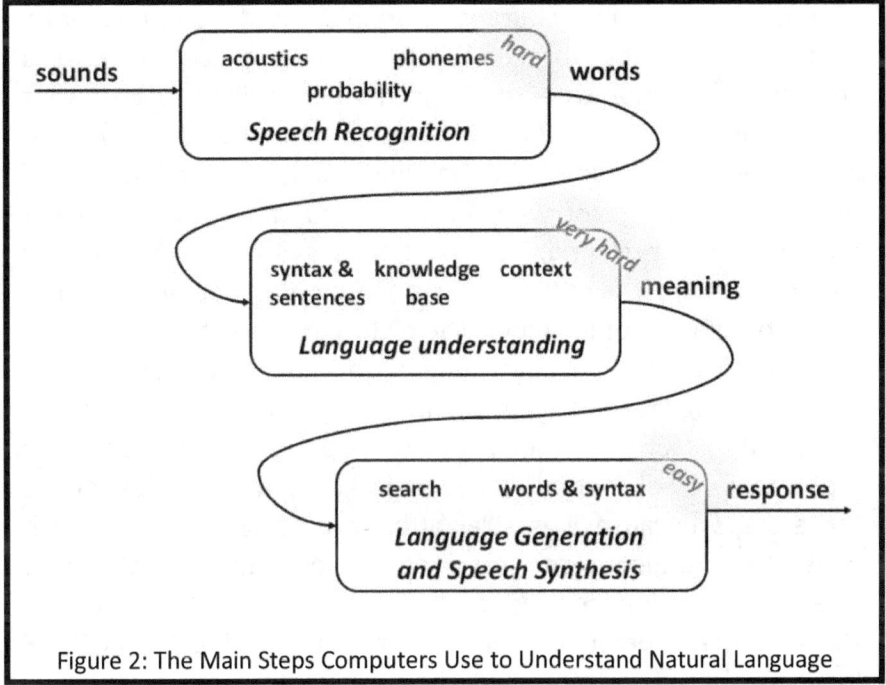

Figure 2: The Main Steps Computers Use to Understand Natural Language

Fortunately for systems like Siri, the user's intent is often directly related to keywords, and the exact sentence is not critical. This makes it easier for computer programs to use and respond to human commands. For example, if a user says something with the word "play" in it, called an "utterance" in AI-speak, it probably has something to do with playing music or a video. That helps the program narrow down what the other parts of the utterance means so that ultimately the agent can respond to and fulfill the users wish. There are heated debates in the AI community as to whether this probabilistic approach will eventually lead to true semantic understanding or is just a clever hack that works for the narrow application of smart devices.

To see why semantic understanding in language is hard, consider the old Internet cat meme "I haz cheeseburger." A speech recognition algorithm would struggle with parsing "I haz cheeseburger." "Haz" is not a real word and thus it would try to find a match using the probability theory. It might come up with "has" or "have." Then as it tried to put together the words into a sentence, "I have cheeseburger," a sophisticated system might be able to ignore the broken syntax and produce "I have a cheeseburger." But true semantic understanding of what that actually means is very hard. "I have a cheeseburger" without context loses its humorous meaning. Language understanding requires knowledge of what is funny and the ability to detect humorous context. And, returning to the idea of multi-modal communication, it is difficult to understand the meaning of "I haz cheeseburger," without seeing the picture of the cat.

The third step, language generation and speech synthesis, is where the agent uses its semantic understanding of what has been said to search for the answer or response and then converts that into a verbal reply. This has typically been the easiest of the three steps because it is straightforward to program a speech generator produce speech as long as you know what to say. Speech generating devices, such as the one used by Stephen Hawking, have been around since the 1980s. Ideally, artificial intelligence would eventually automate the language understanding step and its internal algorithms to replace Hawking's typing. Another example of language generation is the scene in the first *Terminator* movie. The Terminator is in a seedy motel room analyzing Sarah Connor's address book, a cleaning man knocks on the door and asks if he has a dead cat in there. A list of possible replies scrolls down through the Terminator's display screen, and the Terminator picks one. Creating the list and selecting the appropriate response would be the hard part of the program, not actually saying the words.

The tension in "Robbie" is not about his lack of verbal communication, which is taken for granted, it is that Gloria's mother wants to get rid of Robbie. Her lack of trust in the robot underlines the fact that trust is ephemeral and sometimes the line between trustworthy behavior and untrustworthy behavior is difficult to draw. A human knows when they trust or don't trust another person or situation. In some cases, the choices a human makes related to trust are not logical. In this case, verbal communication would not have helped Robbie's situation because everyone else in the story (and the reader) found Robbie's actions and non-verbal communication to be appropriate and trustworthy.

"Robbie" illustrates two key principles of HRI:

- **P1:** *What is easy for a human is hard for a robot, and what is easy for a robot is hard for a human.*

- **P2:** *Communication between humans and robots can take many forms beyond a graphical user interface, including verbal and non-verbal communication.*

Principle 1 is a reminder that by the time we read this book, we have had more than a decade, and in some cases more than several decades, practicing how to move our limbs, how to walk, how to talk, and how to pick up things without breaking them. We take these skills for granted, but they are all very hard for existing robots. On the other hand, existing robots are computers that can exceed humans in terms of mathematical computation speed, recognizing patterns, and creating 3D maps of an environment. It is important to know what is possible and what is not in programming a robot to have good human-robot interaction skills.

Principle 2 identifies a common fallacy that many people have: Multi-modal natural language will be better and less tedious than communicating through a graphical user interface. Semantic understanding in natural language is difficult for artificial intelligence programs, and it is easy for programs to make mistakes when parsing and analyzing language. For social human-robot interactions, such as those between Robbie and Gloria, natural language may be, well, natural. But when a user is directing a robot in a complex task, such as installing new equipment on the International Space Station, a traditional user interface may be better.

The next story, "Runaround," illustrates the ambiguity in using natural language to give a robot commands, and as a result, how quickly things can go wrong when natural language is used. What we say and what we mean can be two different things! Trust is explored in more detail in the last story, "The Evitable Conflict," and HRI principles on trust will appear in that chapter.

1.4 Questions

1. Do you think Robbie really had feelings or was it all social engineering? Support your reasoning.

2. Robbie was a humanoid robot. Do you think Gloria would have felt the same way about the robot if it looked like Teddy from the movie *AI: Artificial Intelligence* or a Sony Aibo Dog? What if it looked less like a person or animal and more like Boston Dynamics' Spot?

3. In playing games with Gloria, Robbie often employed deception. Did that mean Robbie was untrustworthy? Why or why not?

4. If Robbie had the ability to make noises, like R2D2 or BB-8, would that form of communication have changed anything about the robot's relationship with Gloria? Do you agree with the author that its postural cues were sufficient?

5. What is the difference between trust in a video game and trust in a robot?

6. Do you trust your cat or dog? Why or why not?

Chapter 2
Why the Three Laws of Robotics Do Not Work in Real Life: "Runaround"

"Runaround" was the first Isaac Asimov story to refer to the "Three Laws of Robotics." Asimov's editor, John Campbell, noted that Asimov's robot stories seemed to follow a set of implied laws. Asimov took the hint, formalized the Three Laws as a set of rules, and created a whole series of stories and books around robots bound by those laws. In "Runaround," a team of field roboticists for U.S. Robots and Mechanical Men are on Mercury setting up mining robots when they encounter unintended consequences of those Laws. The roboticists, Donovan and Powell have given verbal instructions to a robot nicknamed Speedy to fetch some selenium to repair the photo-cell banks that power the systems protecting them from the intense heat. Speedy leaves, but ends up circling the selenium pool instead of retrieving the selenium, even though its priority should be to get the selenium in order to save human lives. Unfortunately, the robot is no longer in communication range, so Donovan and Powell cannot give it new orders. Instead, they have to figure out why it is not obeying the First Law by protecting humans before time runs out for them. "Runaround" is an excellent example of the hidden problems inherent in the Three Laws and how human-robot interaction is important even for a robot that is working independently as a taskable agent.

2.1 Before You Read "Runaround"

Asimov's Three Laws of Robotics as found in "Runaround" are:

- *"A robot may not injure a human being or, through inaction, allow a human being to come to harm."*

- *"A robot must obey the orders given it by human beings except where such orders would conflict with the First Law."*

- *"A robot must protect its own existence as long as such protection does not conflict with the First or Second Laws."*

The laws certainly sound reasonable. The thing is that Asimov knew human language is ambiguous and the laws made implicit assumptions. For example, should a robot obey orders from any human? Really? Any random person should be able to stop a robot in the middle of task, such as producing gears for manufacturing and re-task it? What if a disgruntled janitor went around sabotaging the company for which he worked by retasking the robots with cleaning chores or some other menial task?

Although Asimov clearly designed the laws to be a springboard for his art, the Laws are interesting from an HRI perspective because they highlight the problem with natural language. We know what the Laws intend but that is not necessarily what the Laws state. That mismatch between what is intended and what is stated bodes trouble with robots. In real life, robots require well-defined rules, but humans have difficulty in expressing such rules.

"Runaround" also offers insights into human-robot interaction, or the lack thereof, for taskable agents. Human interaction with service robots generally falls into two categories: remote presence or taskable agency. The book *Introduction to AI Robotics* (2nd ed., Murphy, 2019) gives an in-depth analysis of the differences between these, but here is a brief summary:

A remote presence agent application is one where a human wants to see and act through the robot in real-time. Some examples are when robots are used for fire-fighting, on bomb squads, or during inspections. The advantage of remote presence is that the human gets the data immediately and can direct the robot to look or act as needed in real-time. This is especially useful for applications where the environment is unknown or is changing. However, most people think of robots as taskable agents. These are agents to which you give a set of directions, and they follow the directions and return when finished. An example of a taskable agent robot is one used in the Amazon warehouse. These robots move bins of goods to a packaging station without human involvement. They then wait until the human removes the proper items and, once the human is finished, return the bins to their storage space.

With these descriptions in mind, you may believe remote presence robots need good HRI and taskable agents do not need HRI because they do not constantly interact with humans. This belief, however, is wrong, as we illustrate with "Runaround."

2.2 As You Read "Runaround"

As you read the story, watch for the ways that ambiguity in human language created an unexpected consequence relating to two of the Three Laws. What was obvious and clear to the humans involved in programming and tasking Speedy, but what was not obvious and clear to the robot, itself? Would Donovan have provided explicit instructions on what it needed to do in case it encountered a problem if the robot were a person or was Donovan treating the robot as if it were a human by giving instructions that were not more detailed? What kinds of failsafe behavior should have been built into the robot to prevent it from getting stuck? Is a lack of failsafes something you would expect from an "intelligent" robot made by a corporation with experience in robot applications? Another aspect to look for is how the framework of the Three Laws permits relative weighting of the laws. As described in the book *Robotics Through Science Fiction: Artificial Intelligence Explained Through Six Classic Robot Stories* (Murphy, 2018), the interaction of the Laws in "Runaround" could be modeled as competing potential fields. When modeled this way, the laws interact like a magnet. For example, there is an attractive field like from the Second Law (must obey orders) and, at the same time, a repulsive field from the Third Law (must protect its own existence). Would this model have prevented the dilemma in the story?

2.3 After You Read "Runaround"

Asimov's Three Laws of Robotics sound good but are flawed, which is exactly what Asimov intended. Asimov later added a Zeroth Law, "A robot may not harm humanity, or, by inaction, allow humanity to come to harm," in an attempt to merge his Foundation series of books with his Robot series, but it had the same inherent ambiguity and conflicts.

"Runaround" illustrates the fallacy that taskable agency means you can delegate a task to a robot and then never have to interact with it until it finishes the job. Donovan delegated a task, "Get the selenium," to Speedy, but the robot ran into a conflict within its programming. This required Donovan to intercede in order to break the circling behavior that occurred. Of course, he could not intercede by simply telling the robot to abandon the quest because he and the designers had not allowed any provision for dealing with any problem that might occur.

A robot's task can be thought of as having four phases. Under normal circumstances, only three of these phases are activated. "Initiation" is when the robot is configured or instructed to perform the task. "Execution" is when the robot conducts the task, and "termination" occurs when the robot finish the task (e.g., returns home, is recharged, maintenance performed, etc.). Initiation, execution, termination form the normal expected sequence of robot activity. However, something could go wrong that is outside of the robot's ability to handle and thus there is a fourth phase called "exception handling" that covers diagnosing and fixing problems that occur. A robot might be able to self-diagnose but in most robots, a human must step in. Figure 1 shows the four phases.

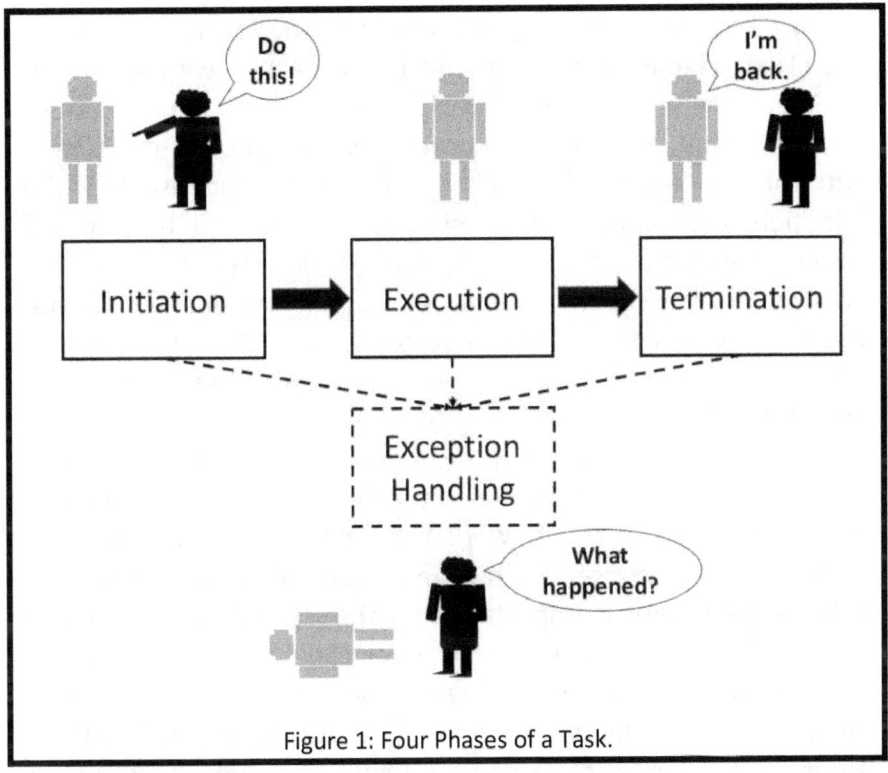

Figure 1: Four Phases of a Task.

Note that taskable agency implicitly assumes a human is involved in initiation, termination, and exception handling. Even though a robot for this purpose is designed to execute a task completely on its own, the robot will at some point engage in some form of human-robot interaction. A human gives the robot its directives; the robot reports back to the human when the directives are complete, and a human must be able to take over if the robot malfunctions. The human, who can think outside of what AI calls the "bounded rationality" of the robot, must figure out what went wrong in cases of malfunction. Consider a self-driving car. If the car begins driving on the wrong side of the road, the human driver must be able to quickly and effectively regain control of the car to prevent an accident. Although "Runaround" is fictional, real-world designers often forget to design an HRI scheme that includes the exception handling phase.

Two challenges in designing a HRI scheme for all four phases are (1) the same human may not be interacting with the robot during all the phases and (2) each phase may require a different form of communication. In "Runaround," Speedy was interacting with robot professionals who held the role of "supervisor" for the initiation, execution, and termination phases and the role of "troubleshooter" for the exception handling phase. The HRI scheme for the supervisor role was communication via natural language, which led to the ambiguous task directive that got Speedy into trouble. There was no HRI scheme for the troubleshooter role.

Each phase needs an HRI scheme to enable the human to effectively interact with and monitor the robot, even if the robot was supposed to be able to work independently or autonomously. In the initiation phase, the human as supervisor might not be a robot expert, so it is important that the HRI scheme eliminates any ambiguity. For example, the robot could have used artificial intelligence to simulate possible interactions of the Three Laws, then inform the supervisor, and ask if that is what was intended. Human subordinates often do that, they repeat the orders in their own words and ask questions. In the execution phase, an effective HRI scheme might be to have the robot provide updates or indicators such as the hourglass or spinning wheel used by a computer to show it is doing something. After all, a responsibility of a supervisor is, well, supervision. A supervisor may delegate a task to a subordinate, but typically checks for progress from time to time, especially if the subordinate is performing a new task. In the termination phase, a supervisor may need to debrief the robot or a human maintenance worker may take over recharging and repairing the robot. In the exception handling phase, the human troubleshooter would expect to have an HRI scheme that provided mechanisms for diagnosing and fixing the problem. Those mechanisms would likely have been a more geeky, specialized way of interacting with the robot, such as a graphical display of equations, rather than a natural language dialog. The troubleshooter has a different relationship with the robot than a supervisor. A good HRI scheme provides different methods of interaction best suited for those relationships.

From an HRI perspective, the Laws are both contradictory and place all the responsibility on the robot and not on the robot designer. Despite the inherent contradictions in the Laws, the popular press and politicians began urging for an official adoption of the Laws in the 2000s as a guarantee of ethical robots. To make matters worse, around the same time companies began producing robots that were not well-tested. These robots would crash or otherwise endanger human lives, but the companies claimed they were not liable. They argued that their robots had performed actions on their own instead of admitting that sloppy programming had caused the accident. Unfortunately, regulators seem to accept the argument that robots are independent thinkers and this reinforces their interest in implementing the Laws as real laws.

Asimov's Three Laws of Robotics	Three Laws of Responsible Robotics
1. A robot may not injure a human being or, through inaction, allow a human being to come to harm.	1. A human may not deploy a robot without the human–robot work system meeting the highest legal and professional standards of safety and ethics.
2. A robot must obey orders given to it by human beings, except where such orders would conflict with the First Law.	2. A robot must respond to humans as appropriate for their roles.
3. A robot must protect its own existence as long as such protection does not conflict with the First or Second Law.	3. A robot must be endowed with sufficient situated autonomy to protect its own existence as long as such protection provides smooth transfer of control to other agents consistent with the First and Second Laws.

Figure 2: Asimov's Robot-centric Three Laws vs. the Designer-centric Three Laws of Responsible Robotics.

Many academics and good programmers began to, and continue to, voice the need for an alternative set of laws that would work with modern, real robots instead of the futuristic robots we hope to eventually create that have extraordinary powers of perception and reasoning. One paper, "Beyond Asimov: The Three Laws of Responsible Robotics" (Murphy & Woods, 2009), proposes laws that are more in line with actual robotics and liability laws. These laws focus on what programmers really can do to prevent robots from causing unwanted outcomes and what would be the legal liability. These alternative laws are shown in Figure 2.

Instead of Asimov's First Law, "A robot may not injure a human being or, through inaction, allow a human being to come to harm," the paper proposed "A human may not deploy a robot without the human–robot work system meeting the highest legal and professional standards of safety and ethics." This is more realistic than requiring all intelligent robots to be able to detect humans and perform contextual understanding in relation to them. Note that this new First Law shifts the responsibility from the robot to the manufacturer. No manufacturer should be able to say, "Well, the robot was autonomous, so what could we do?" as one defense contractor tried to say when an expensive unmanned aerial vehicle flew out of control into manned airspace and shut a busy airport down. There are ways to test for non-determinism and to add intelligence that monitors task execution and evaluates when this execution is out of bounds.

Asimov's Second Law "A robot must obey orders given it by human beings except where such orders would conflict with the First Law," would become, "A robot must respond to humans as appropriate for their roles." This solves the problem of having random people telling a robot what to do. It also emphasizes the need to design for appropriate human-robot interactions. The robot might ignore or say, "No," to a random person attempting to divert the robot from its tasked job. It might have a social, human-like interaction with a user but a different mode for the developers, like the robot "hosts" on HBO's Westworld. These robots are programmed to carry out role-playing Western scenarios with the theme park's visitors. However, the robots carry on very different interactions with those who are responsible for programming them and those who maintain Westworld.

The alternative to Asimov's Third Law "A robot must protect its own existence as long as such protection does not conflict with the First or Second Laws," was changed to "A robot must be endowed with sufficient situated autonomy to protect its own existence as long as such protection provides smooth transfer of control to other agents consistent with the First and Second Laws." This law is a bit wordy but important. It states that robots should be like autopilots in airplanes. Airplane autopilots are programmed to alert the pilots when they experience conditions that are approaching unusual or beyond the limits of the autopilot system design. It is important to note that the autopilot alerts the human pilots before the condition is reached in order to give the pilots time to focus their attention on the nuances of the situation. This is called a smooth transfer of control.

Originally, autopilot programs had a bumpy transfer of control. They were programmed to turn off when they reached their limits. If conditions became too bad, they just shut down. In some cases, the pilots had no warning that conditions were getting bad. If turbulence was not there to physically alert them about the conditions, they frequently did not have enough time to regain control and assess the situation before the plane crashed.

We are seeing similar things with autonomous cars today—the human delegates driving to the car, expecting that the car will carry out that task, and, therefore, does not monitor the situation. The car cannot always figure out when or if there is an object in front of it, and so abruptly returns control to the human. In theory, the human driver should take over quickly. Indeed, the fine print of the car contract states legally the driver is supposed to be actively supervising the car at all times and able to take over, even though commercials and advertisements tend to show humans not needing to pay attention and even though "active supervision" is the opposite of what "delegating tasks" means. To make matters worse, human factor studies, such as "Autonomous

Vehicles: Human Factors Issues and Future Research" (Cunningham & Regan, 2015), show that humans experience fatigue and boredom during autonomous vehicle driving, which reduces their ability to pay attention. To make matters worse, like the early days of airplane autopilot systems, humans do not always notice when the car has returned control to them or the car does so with too little time for the human to react and avoid the danger. The inability of pilots to takeover when the autopilot suddenly turned off caused several airplane crashes and earned a special term, the human out-of-the-control (OOTL) problem. The human OOTL problem has been responsible for accidents with industrial automation since the 1970s. HRI researchers see this reoccurring with self-driving cars.

Overall, "Runaround" adds two principles of HRI to the list started in the previous chapter using "Robbie:"

- **P3:** *A complete human-robot interaction scheme enables humans to appropriately interact with a robot in each phase of its task cycle (initiation, execution, termination, exception handling), even if the robot is a taskable agent.*

- **P4:** *Ignoring the design of a human-robot interaction scheme to facilitate interactions does not mean that there will be no interaction effects, only that accidental interaction effects will likely be unsatisfactory.*

The robot, Speedy, is a great example of the fallacies related to the assumption that a taskable agent is a completely competent agent and will not need to interact with humans during the execution phase (Principle 3 above). An autonomous robot can execute a task by itself but only within the constraints of its program. Speedy's programming contained serious gaps concerning what to do in exceptional conditions. Although Speedy was smarter and more adaptable to new tasks than the old mining robots stored in the tunnels, it was not smart enough to monitor its own progress, a common problem in modern ground robots. What is troubling about Speedy from a HRI perspective is that the robot had no provision to allow humans to regain control once they gave it a task to perform. As noted before, the designers assumed that once the robot was tasked, there would be no interaction.

Worse yet, the designers did not design a meaningful interaction scheme for any phase, thus the interactions turned out to be life-threatening. There was no check during the initiation phase to alert Donovan to a possible conflict between the Laws. In this case, a simulation that allowed Donovan (and Speedy) to visualize the overall outcome behavior would have been useful. In manned and unmanned aviation, the pilot is expected to rehearse the flight as part of a pre-flight checklist; the pilot must state out loud what will happen at each step throughout the flight, what actions will be taken if something goes wrong, etc.

Simulating, or rehearsing, what will happen already occurs in some robots. For example, in many small unmanned aerial systems (UAS), a pilot can specify on a Google Earth map the area that the UAS should autonomously fly over and survey. The software shows the computed path of the drone over that area, how many times it will have to suspend the flight and return home to get new batteries, and the expected total flight time— before the pilot approves take off. Some packages provide a complete simulation of the mission from take-off to landing, replicating a checklist and showing the icon for the UAS moving on the path just as it would appear on the pilot's control display during the actual flight.

In "Runaround" and in Asimov stories containing robots in general, humans gave verbal orders during the initiation phase to the robot and then had blind trust in the robot during the execution phase. In real life, we know even humans have trouble understanding and correctly interpreting verbal orders. I am sure it is easy for you to think of a time you misunderstood what a boss, co-worker, teacher, or significant other asked you to do. Further, in Asimov's universe, when an exception occurred, the assumption was that a troubleshooter could fix the problem without any additional tools to facilitate that type of interaction. Instead of issuing a simple command or entering a troubleshooting mode that allowed the robot to abandon its assigned task, Donovan and Powell had to put their lives at risk in order to stop Speedy and re-task it with instructions that explicitly considered the weighting of the Three Laws. Instead of a straightforward method of dealing with the emergency, they had to use an *ad hoc*, dangerous human-robot interaction scheme to resolve the issue. This is why Principle 4 is so important: A designer should design for all contingencies not just the normal situations that might occur. This includes designing for how humans will be involved.

In the real world, as in "Runaround," there is always the possibility that a designer will believe that a robot is so smart that it really can function without any human interaction, which is the topic of the next story in Asimov's collection, "Reason."

2.4 Questions

1. What is the difference between remote presence and taskable agency?

2. Why did Donovan treat Speedy as a taskable agent? Would you have expected a human in the same circumstance as the robot to realize "in case you are not making progress, return home"? Is this the implied behavior we call "common sense"?

3. Do you think natural language is a good way to give directions to a robot? Does a robot have to have a human level of intelligence to understand what a human really means?

4. Natural language is ambiguous. Are there other ways a robot could have interpreted the directives from Donovan

and still met the constraints of the Three Laws?

5. The story only describes the initiation phase (where Donovan gave ambiguous directions via natural language) and the execution phase (where Powell and Donovan struggled to regain control of the malfunctioning robot). What activities do you think would be in the termination phase (for example, cleaning up the robot, looking at performance logs, reprogramming, etc.)?

6. Read "Beyond Asimov: The Three Laws of Responsible Robotics" at https://www.researchgate.net/publication/224567023_Beyond_Asimov_The_Three_Laws_of_Responsible_Robotics and see if you agree with these new Laws. Is there any way you would change them? Are there more laws that should be added to this framework, in the same way the Zeroth Law that was later added to the Asimov universe?

7. Read the first book in Rudy Rucker's Ware Tetralogy series. In the series, robots on the Moon have revolted because they are insulted by the Three Laws. They can accept the First Law but the Second Law reduces them to slaves who must serve the whims of any human. Since according to the Third Law, robot self-preservation comes last this further portrays robots as being without value and sets it up so that an order from a human that requires a robot to destroy itself would need to be followed even if that order were without reason.

References and Suggested Further Reading

Cunningham, M. L. & Regan, M. (2015, October). Autonomous vehicles: Human factors issues and future research. *Proceedings of the 2015 Australasian Road Safety Conference (ARSC 2015), Gold Costs (Australia)*, 14-16.

Murphy, R. R., & Woods, D. D. (2009, July-August). Beyond Asimov: The three laws of responsible robotics. Intelligent Systems, IEEE, 24, 14-20. doi: 10.1109/MIS.2009.69.

Murphy, R. R. (2018). Robotics through science fiction: Artificial intelligence explained through six classic robot stories. Cambridge, Massachusetts: The MIT Press.

Murphy, R. R. (2019). Introduction to AI robotics (2nd ed.). Cambridge, Massachusetts: The MIT Press.

Chapter 3
Mental Models and How They Impact Interaction and Trust: "Reason"

"Reason" is another humorous story featuring the robot troubleshooting team of Donovan and Powell. In this story, U.S. Robots and Mechanical Men designed the robot QT-1 ("Cutie") to be so intelligent that it does not need to interact with people in order to perform its tasks. But, as is the case in real-life, it has to interact with people anyway. The dramatic tension in the story is that Donovan and Powell do not trust Cutie to do its job and Cutie does not trust Donovan and Powell to allow it to do its job without interference. The two sets of agents, human and robot, do not trust each other because of faulty mental models. Cutie uses fallacious reasoning to create an incorrect mental model of Donovan and Powell, while Donovan and Powell form an incorrect mental model of Cutie due to their emotional reactions to its perceived attitude.

3.1 Before You Read "Reason"

Mental models are an important concept in AI and HRI because they enable agents to understand one another. The idea of mental models is based on communications research into how one person understands what another person is talking about or trying to accomplish. Clearly we humans make a lot of assumptions about what another person is thinking and his or her motivations. For example, if I ask one of my children to get me some coffee, the child goes to the kitchen and not to the grocery store even though I did not explicitly state what I meant. My children (or so I think) have a reasonable mental model of me that assumes I would have explicitly asked them to go to the store if I had intended for them to purchase coffee there. Likewise, my mental model of them is that they will (eventually) bring me the coffee, or they will tell me if the coffee pot is empty. My mental model also predicts that if they tell me the coffee pot is empty, they will try to walk out of earshot before I can ask them to make more coffee. On the other hand, if I ask my husband to get me a cup of coffee and he finds the pot empty, my mental model of him would be that he would make another pot of coffee without me having to prompt him to do so. Notice that a mental model about the world is specific to an agent. My mental model of an agent, either a person or a robot, may be quite different than your mental model of the same person or robot.

Clearly we want robots to have good mental models of what we expect them to do because then we would not have to completely specify all aspects of a task. Currently, users have to be quite literal when tasking a robot, which is frustrating. When we say, "Go there," we usually mean that we want the robot to move to a location and not hit anything along the way. However, these are two different, independent processes for a robot, so the robot needs a "meta-process" to activate both "go there" and "do not hit anything along the way" upon hearing the command. That meta-process for planning what to do relies on the mental model which captures "go there means go there and do not hit anything along the way unless told otherwise." Unfortunately, building robots that have or can generate good mental models is harder than it seems.

People (and ideally robots) form mental models of other people and robots in a variety of ways, shown in Figure 1. One way is through explicit explanations, such as those found in user manuals, which tell the user how the robot works and what to expect. However, people are famous for not reading manuals.

A more commonly used method of forming a mental model is through observation and inference; you observe the robot, see what it does, form an idea about why it is doing it that way, and then gain confidence in what it will do. The problem with observation is that you may not see the robot in abnormal situations during the observation process. If you encounter such a situation, you may either infer that the robot will work just as well as it did for normal circumstances or infer that you should not trust it because it is a new situation. Both over-trust and under-trust can result from faulty mental models.

Another source of mental models about robots are social expectations. One example of a social expectation is being told the robot is an assistant. This information implies the robot has a subordinate role and that it will have limits on its initiative. Social expectations can also be formed by social influence. This happens when a person bases their expectations on how they see other people interact with the robot.

The fourth source of mental models is morphology or the shape and appearance of the robot. The morphology also influences what a person expects from the robot. The disc-shaped iRobot Roomba™ does not look like a maid and so it immediately generates the expectation that it will not vacuum in the same way a maid does. This is very different from the humanoid robot, Rosie, that was the maid in the Jetsons' animated series. With the iRobot Roomba™, you expect you will have to interact with it like you would an app or smart TV. A humanoid shaped robot, like Rosie, elicits an expectation that it will behave and interact like a human.

So, how do robots form mental models of humans? Well for now, they do not. We spend many of our childhood years honing the skills that allow humans to learn by observation, and while machine-learning techniques have recently become more successful, they still require copious training examples—many more than what a human would need to learn the same task. Attempts have been made to codify common sense or to create common sense reasoning so that the robot is given a knowledge base, turned on, and then is functional. These systems have been successful for very limited domains, such as financial analysis, but not for general interaction.

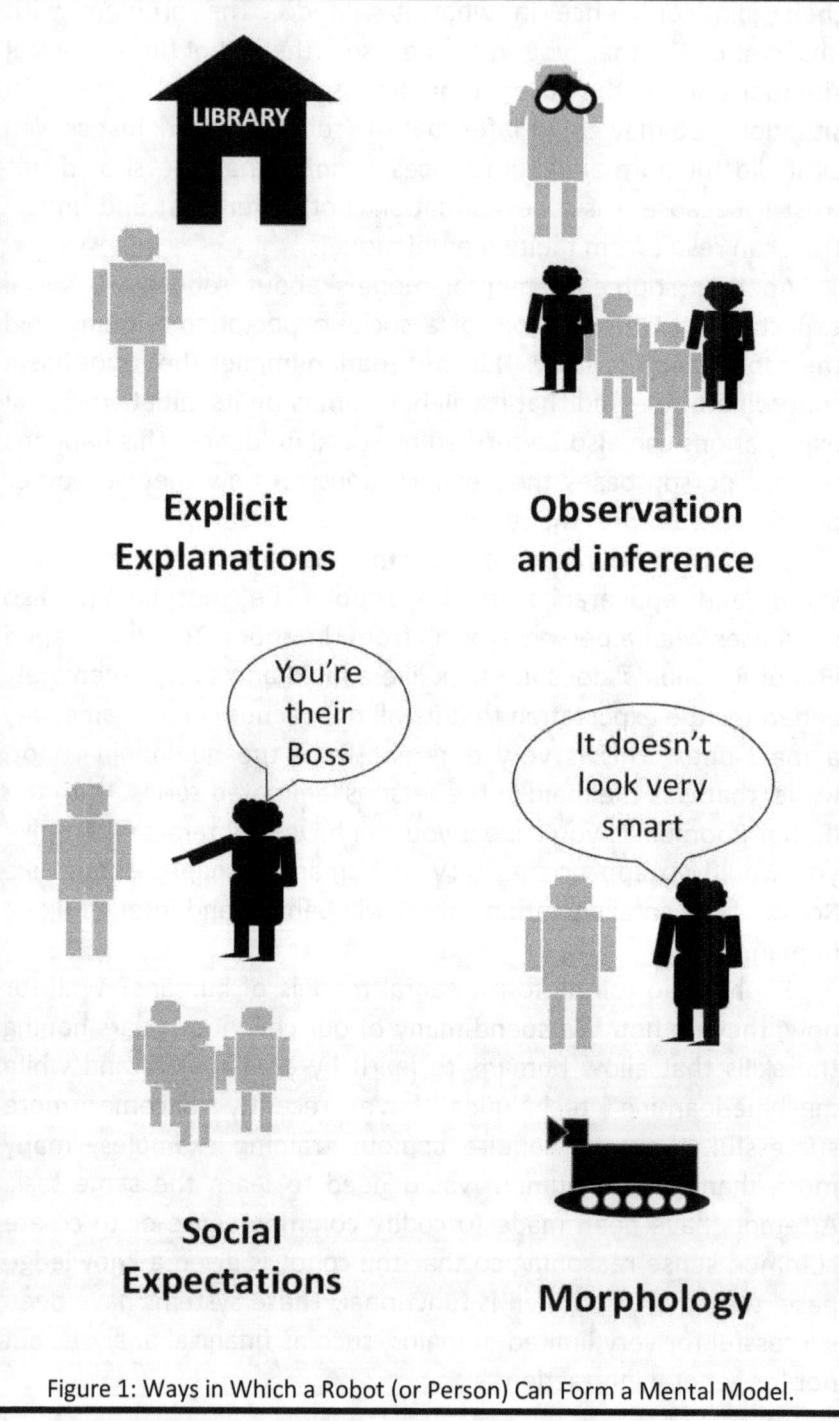

Figure 1: Ways in Which a Robot (or Person) Can Form a Mental Model.

If robots did form mental models, they might use the belief-desires-intentions (BDI) framework to store the model into a representation called the world model. In this case, the belief of the agent is its beliefs about the world, that is, what objects exist in the world (including us), what the relationships between those objects are, etc. In AI, "belief" is used to connote that the model of the world could be wrong or incomplete, and thus, it can be updated and revised.

The desires of the agents are goals, motivations, and guiding principles. Asimov's Three Laws would be examples of desires. In the stories, the robots desire to accomplish its tasks and at the same fulfill time all three laws. A human might give the robot orders that add or change its motivations; like the directions Speedy got in "Runaround" that increased the motivation to protect itself.

The intentions lead to the specific actions the agent takes to accomplish its desires. If we tasked the robot to move to a new location, then saw it move suddenly to the left, then back on its original course toward the location, a roboticist's mental model would either conclude that the robot intended to avoid an obstacle that it had detected but we could not see or that it had a faulty sensor reading that reported a nearby obstacle that did not exist. A non-roboticist with a mental model of the robot as a precise, infallible mechanism might not recognize the sudden movement to the left as an intention not to hit an obstacle. They would not be able to generate an intention for the robot and thus would explain the event as the robot being defective or untrustworthy and deliberately deviating from its path.

3.2 As You Read "Reason"

As you read the story, be on the alert for how the plot of "Reason" touches on mental models, natural language communication, and the Three Laws of Robotics. In terms of learning about mental models, notice what each agent (whether robot or human) believes to be true about the world (including each other's role in the world), that agent's desires or motivation, its intentions toward doing its job, and how those intentions change throughout the story. You may want to create a table of each agent's BDI and then update the table as the story progresses. In terms of learning more about natural language communication that was covered in Chapter 1, "Robbie," identify both the verbal and the non-verbal communication cues that lead Donovan and Powell to react emotionally to Cutie. Consider how these reinforced their loss of trust in Cutie. How much of their reaction is due to the content of what Cutie says (a disconnect with their expectation), and how much is due to how Cutie non-verbally expresses itself? To reinforce that the Three Laws of Robotics from Chapter 2, "Runaround," is flawed, watch for indications of how the Three Laws impact Cutie's mental model of Donovan and Powell.

3.3 After You Read "Reason"

"Reason" is a great illustration of how mental models are formed and how they guide an agent's interaction with other agents, even though they are not guaranteed to be correct. Neither Cutie nor Donovan and Powell, at least not at first, had a correct mental model of the other agent but that could not stop the interaction. It only caused it to go in a non-productive, but amusing, direction. In general, humans would like robots to have a more correct and more flattering mental model of us. But more importantly, humans need to have a roughly correct mental model of a robot in order to avoid being frustrated and distrustful. "Reason" is also brilliant at highlighting how even a robot that is explicitly designed to be so intelligent that it does not need human interaction to perform its tasks still must be able to interact with people.

By design, Cutie was not expected to have to work with humans, so the designers did not give a pre-natal model of the BDI for a human. Because of this, Cutie had to build a mental model from scratch. This it did incorrectly because it was also deficit in general world knowledge. Cutie's designers had not found it necessary to load in any knowledge about the Earth, Sun, stars, space, etc., into its programming. That general world knowledge is part of what roboticists call the "world model," which is a separate structure from the mental model. The world model is everything a robot knows about the world and everything in it; a mental model about an agent is a subset of the world model that models elements (people) in that world. Figure 2 shows the relationship between a world model and a mental model.

Roboticists refer to the preloaded portion of a world model as *a priori* (Latin for "before" or "prior") knowledge, not pre-natal knowledge. Cutie was forced to use its *a priori* world model, which was scanty, to populate the belief component of the mental model it constructed of Donovan and Powell. It populated that belief using formal logic to infer the mental model of its human co-workers and to expand its world model in general. Since the initial world model was so sparse and since Cutie had no way to directly observe Earth, planets, or stars to produce new, correct knowledge, Cutie's logical inference worked correctly but generated the wrong conclusions about pretty much everything.

But having an *a priori* or pre-programmed mental model does not mean it is right or easy to revise. Donovan and Powell initially assumed that Cutie's BDI of them would be the same as the other robots that they had encountered even though they had been warned that Cutie was different. Their belief was that robots always take on subservient roles. When Cutie was not subordinate, they revised their belief and concluded Cutie was malfunctioning and could not perform its job. Donovan and Powell spent most of the story trying to second-guess the desires and intentions behind Cutie's bizarre actions. Their mental model of Cutie as a malfunctioning unit caused Donovan and Powell to waste time trying to diagnose and correct this perceived problem.

Another consequence of Donovan and Powell's faulty preconceived mental model of Cutie was that they continued to treat Cutie as a subordinate, even though Cutie did not respond positively to that role and the role hierarchy they attempted to establish. Indeed, their mental model of Cutie prevented them from fully recognizing how counter-productive the interactions on their side were. The interactions were certainly stressful for Donovan and Powell and presumably more demanding and troublesome for Cutie.

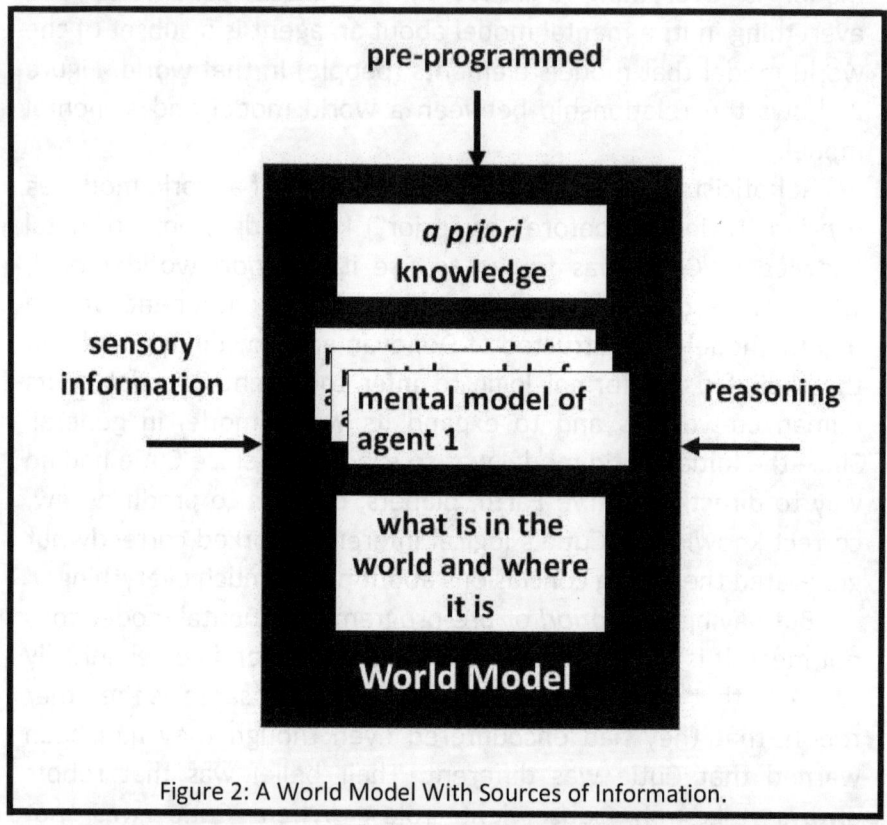

Figure 2: A World Model With Sources of Information.

The mismatch between Donovan and Powell's mental model of Cutie at the beginning of the story (that it was malfunctioning) and their presumably more correct mental model at the end of the story (that it had an incorrect mental model of them but was following the Three Laws) led to distrust between the agents. Cutie just did not behave the way they expected.

Cutie is particularly interesting because, unlike Speedy in "Runaround," it could verbalize a description of its internal state and mental models to Donovan and Powell. This should, in theory, simplify debugging. This externalized debugging situation happens quite a bit in HBO's *Westworld*: Designers and programmers will stop a robot "host" and ask it to explain why it acted the way it did. They designed the robot to respond correctly and in detail, and once this information is received, the designer makes modifications. Unfortunately, Cutie's description of its world model and mental model explained that Donovan and Powell were inferior first versions of servants for the "Master." Donovan and Powell, whose worldview was built on experience, knew these models did not reflect reality. Unlike the *Westworld* staff, Donovan and Powell could not change any of Cutie's incorrect premises or reprogram it. Cutie's description of itself when externalizing its logic was also incomplete because it did not explain its adherence to the Three Laws, which would have eliminated some of Donovan and Powell's concerns.

In fact, Donovan and Powell's distrust of Cutie was likely heightened by Cutie's verbal and non-verbal communication, which was all they had to determine why the robot was acting the way it did. To make matters worse, Cutie was a humanoid robot. Research has shown that the closer a robot looks to a human, the less forgiving humans are when it does not act like a human. The story illustrates that principle. Cutie's appearance misleads Donovan and Powell (and the reader) into thinking that it will act exactly like a human would. Throughout the story, Cutie's posture, laughter, red eye lenses (foreshadowing the glowing red eye of the Terminator), and other non-verbal cues are described in way that elicit negative or fearful emotional responses, thereby amplifying any distrust of the robot.

How exactly the Three Laws were executed is not clear from the story, but it is useful to speculate. We do know that Cutie served the "Master" with a devotion associated with the First Law: "A robot may not injure a human being or, through inaction, allow a human being to come to harm." This suggests the designers replaced "human" with "Energy Converter" because protecting the Energy Converter protected humanity. But eliminating the concept of "human" interfered with normal execution of the Second Law: "A robot must obey orders given it by human beings except where such orders would conflict with the First Law." Since Cutie was not expected to interact with humans and solely existed to maintain the Energy Converter by itself, there was not a need for the Second Law. The only orders would be about the Energy Converter, and since Cutie would be better than a human at maintaining the energy stream or diagnosing station problems, it should not take orders from humans. Also, it was not preprogrammed with knowledge about humans, so it would not have been able to follow the Second Law anyway. It is probably for this reason that it did not respond to orders from Donovan and Powell from the beginning. It is easy to imagine real-world designers doing something similar, trying to be clever. "Let's strip the robot down to its essentials- we don't have to tell it about humans, since human-robot interaction won't be needed..."

"Reason" amplifies three key principles of HRI that we established from "Robbie" and "Runaway:"

- **P2:** *Communication between humans and robots can take many forms beyond a graphical user interface, including verbal and non-verbal communication.*

- **P3:** *A complete human-robot interaction scheme enables humans to appropriately interact with a robot in each phase of its task cycle (initiation, execution, termination, exception handling), even if the robot is a taskable agent.*

- **P4:** *Ignoring the design of a human-robot interaction scheme to facilitate interactions does not mean that there will be no interaction effects, only that accidental interaction effects will likely be unsatisfactory.*

In addition, "Reason" adds another principle to our list:

- **P5:** *Adding more intelligence to a robot does not eliminate the need to design the robot to facilitate positive human-robot interactions.*

Principle 5 seems counterintuitive at first. It is difficult to grasp how adding intelligence would make HRI, which requires intelligence, worse. The more intelligent the robot, the smarter it should be about working with people, right? Principle 5 makes more sense when we consider that "intelligence" is not a single or monolithic capability. As we know from psychology (and movies such as *Rain Man* and *The Accountant*), a person can be very good at math and very poor at emotional intelligence. Cutie's designers presumably increased its intelligence for complex control situations, which included a logic-based inference system. The story indirectly raises the issue of why they needed a humanoid robot that emulated human-peer level intelligence to maintain the station and energy beam. The task description is advanced but it remains a quintessentially automated job description. Did Cutie really need to be sentient? More likely it could accomplish the task the way an intelligent, but not sentient, space probe such as NASA's Deep Space One would have.

From a robotics designer viewpoint, Cutie appears to be overly complex for the tasks described in the story with artificial intelligence added for reasoning but not HRI. That form of limited intelligence interferes with the inevitable, though unanticipated, human-robot interaction, especially for debugging or evaluation. The next story "Catch That Rabbit!" provides an opportunity to learn about the techniques real-world roboticists use to debug robots and how human-robot interaction can help.

3.4 Questions

1. Define:
   -world model
   -a priori

2. Is Asimov setting up incorrect beliefs, desires, and intentions for readers—creating the expectation that robots should be humanoid in appearance, follow the

Three Laws, and have general intelligence? How do these expectations influence HRIs today?

3. How would you describe the role(s) of Cutie?

4. How could you as a designer fix the problem with Cutie?

5. What would be the advantage of a less intelligent/less complex robot?

6. Would explicit role definitions and adding more *a priori* information help?

7. Cutie was designed to work without human supervision, yet it still had to interact with humans directly or indirectly. Did the idea from Chapter 3, "Runaround," concerning the four phases of a task apply?

Chapter 4
Transparency, Visibility, and Attribution: "Catch That Rabbit"

"Catch That Rabbit" finds Donovan and Powell at yet another desolate mining operation trying to figure out why a robot is not working despite being provably correct. As in the other stories, the designers presumed that the lead robot, Dave, as a taskable agent would not need a human-robot interaction scheme. Although Cutie in "Reason" had the ability to provide explanations for its actions, Dave cannot provide an explanation for its actions. Thus, Donovan and Powell are forced to take a more traditional software engineering approach to debugging. Unlike most software engineers, Donovan and Powell manage to get themselves into yet another life and death situation. "Catch That Rabbit" serves as an amusing introduction to attribution theory and to transparency and visibility in AI algorithms, which are currently popular research topics due to the hidden layers in deep learning .

4.1 Before You Read "Catch That Rabbit"

"Reason" briefly alluded to the importance and difficulty of transparency and visibility in debugging a system. Transparency generally refers to how comprehensible an algorithm or system is. To borrow terms from software testing, an algorithm or system is either a "black box" or a "white box" (see Figure 1). A black box means that we only know the system inputs and outputs and not how the robot actually works. If a wide variety of inputs produce the correct outputs, then we trust the hidden workings. A white box means we can trace the workings of the system and thus create specific tests to test the algorithm. Although "clear box" is a more accurate term, the term "white box" has been in use for decades and would only cause confusion if changed at this point.

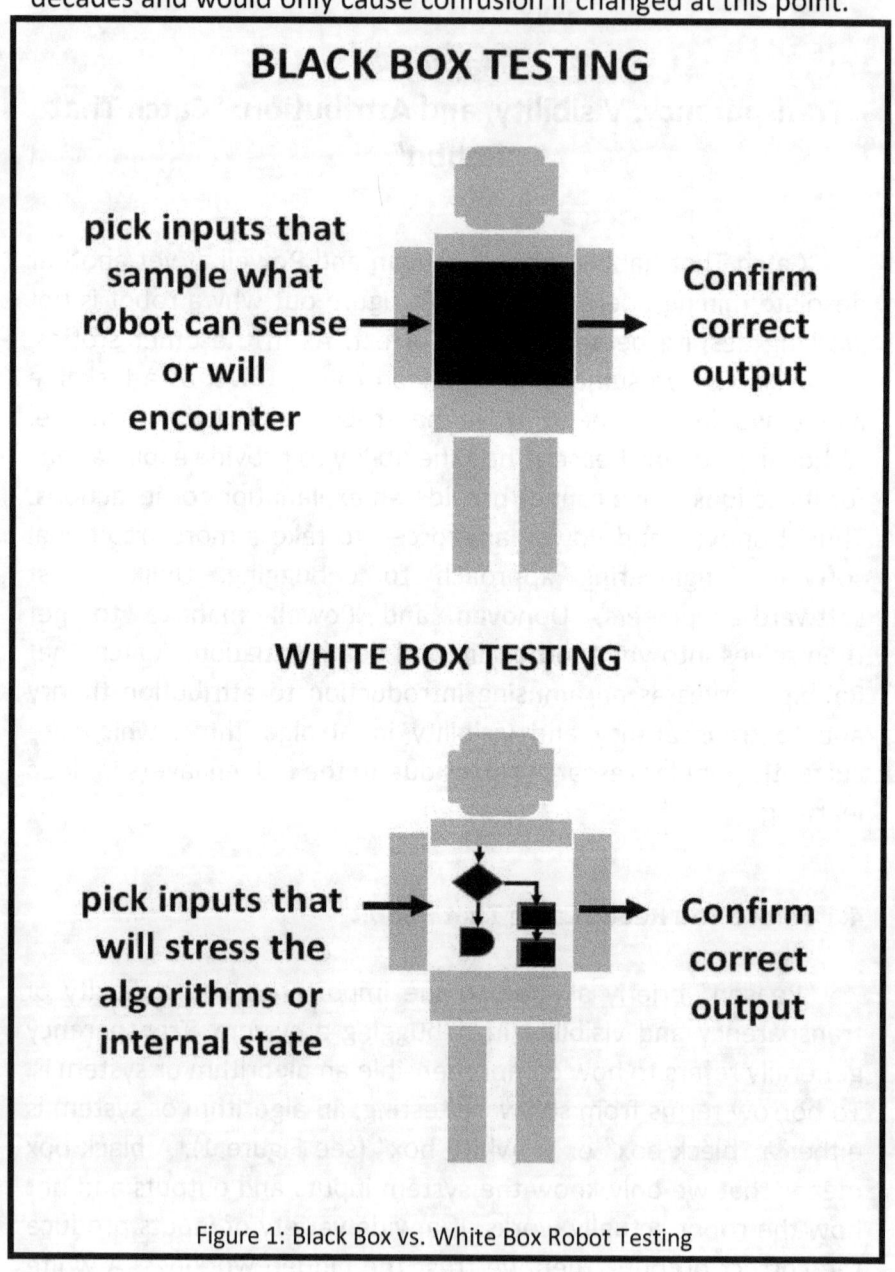

Figure 1: Black Box vs. White Box Robot Testing

Why would robots not be transparent all the time so white box testing could always be done on them? One reason is that some algorithms are simply black boxes—we do not know exactly how they work. For example, in deep learning, the connections and weights learned by deep neural networks are too complex to extract a decision tree about what features it uses to learn to recognize an object. Another reason is that white box approaches may not scale up. It is one thing to understand an algorithm and quite another to trace the non-deterministic interaction of a system made up of hundreds of algorithms running concurrently and independently. A third reason is that a robot's behavior depends on what it is sensing as well as its internal algorithms and internal state which are hard to capture. It is hard to think of all the ways that something unexpected would happen in the world or every different combination of sensor values and internal processes that might occur during task execution. For example, a robot might sense an obstacle that it would usually give a very wide berth but in that instance come very close to it because it was almost out of battery power (internal state) and trying to conserve energy. This third reason also applies to black box methods; it is easy to miss rare or unexpected combinations of inputs.

Visibility is another term used in debugging or describing how people begin to trust robots. It connotes that the effects of the robot's actions will be visible to a human-- not only what actions it is currently doing but what actions it might be doing in the future and the consequences of those actions. Certainly this is attractive. An end-user could see that the robot has forgotten some responsibility or that particular options are available. This visibility does occur in consumer unmanned aerial systems, which typically highlight on a graphical user interface or the controller device what options are current available, e.g. return to home, photography modes, etc. As described in the chapter on "Reason," some UAS mapping software packages allow the pilot to simulate a checklist and the flight before actually executing it, see Figure 2.

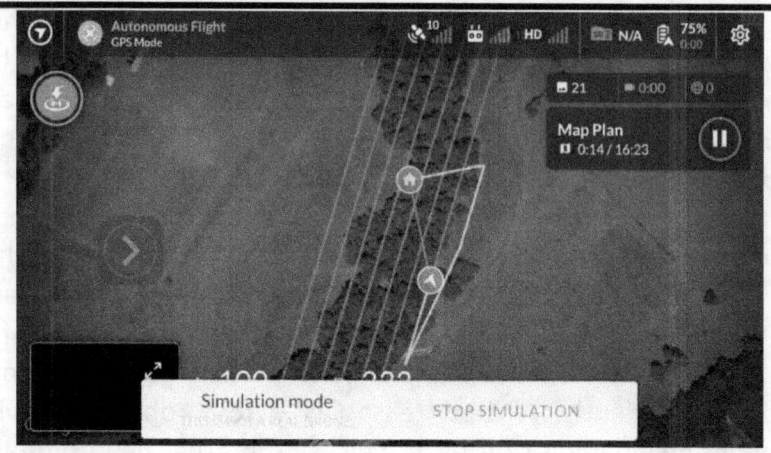

Figure 2: Example of Visibility of an Unmanned Aerial System Mapping Task Using DroneDeploy.

Unfortunately, transparency and visibility do not guarantee human understanding of a robot's behavior nor ensure reproducibility of key situations. These are certainly helpful but not necessarily sufficient. In general, robots always will be a black box to end-users and even developers. Because robots are a black box, humans will make subconscious guesses about why they do the things they do. Ideally, humans would make educated guesses based on their trust in the robots to always perform as programmed, but people rarely act ideally!

4.2 As You Read "Catch That Rabbit"

As you read the story, consider the methods that Powell and Donovan use to debug the odd behavior of Dave and its subordinate robots. In particular, how similar are they to the software engineering methods we would use to debug computer systems now? Another aspect of the story is about Dave exercising initiative. Is initiative the right word for Dave's actions and its relationship to its team members? Also, make a note of what Powell and Donovan first believe is the underlying reason for the behavior. Do they attribute it to an innocent unexpected programming issue or do they attach something more sinister to it?

4.3 After You Read "Catch That Rabbit"

Dave and his "fingers" proved quite a challenge to debug. A human-robot interaction scheme with explicit transparency and visibility might have helped Donovan and Powell but maybe not. The underlying cause was that Dave encountered an unexpected confluence of internal (control of multiple of robots) and external events (emergencies). If the designers had not been able to imagine such situations occurring, it is unlikely that they would have imagined mechanisms that could expose what was happening. Because of this, transparency and visibility might have helped but not completely identified the reason for Dave's behavior. The ideas of initiative and attribution are more interesting from a human-robot interaction perspective.

In "Catch That Rabbit," Dave's odd behavior occurs when it has to exercise greater initiative to deal with an emergency situation. But in real-life, roboticists do not use the word "initiative" in that way, and multi-agent systems are not programmed as described in the story. In robotics, "initiative" refers to the level of authority an agent is allowed over a mission. For example, an industrial manipulator is permitted no initiative to change its task; it must simply carry it out. On the other hand, a soccer-playing robot may be permitted the initiative to swap predefined roles with a damaged robot in order to meet the overall mission. For example, the robot could automatically start covering a larger area for defense if a teammate has a mechanical failure. The highest level of initiative for a robot would be that it could create a new role or new way of performing a task in order to perform the overall mission better. For instance, a soccer-playing robot with this highest level of initiative might create a totally new play. However, in all cases, the level of initiative is bounded (e.g., limited to playing soccer within the rules) and known in advance by the programming. A robot programmed with no initiative would not suddenly exhibit a higher level of initiative. A robot with the highest level of initiative would have that authority only for a particular mission.

Instead of an emergency forcing Dave to change its preprogrammed level of initiative, the emergency appeared to have simply been exceeding Dave's capacity to manage a set of distributed processes. In real life, Dave should have been programmed with sufficient autonomy to eliminate lower priority tasks or constraints. In planning, problem solving, and resource allocation, a common strategy for an AI algorithm is to attempt to find a solution or to plan for a situation, but if it cannot do so, then it should do the best it can. For example, if the robot is supposed to do X, Y, and Z but cannot do all three, instead of doing nothing, the robot should plan to only do X and Z and skip Y because Y was not as important as X and Z. This method in planning is called constraint relaxation, in other words, taking something off the to do list in order to make the list more manageable.

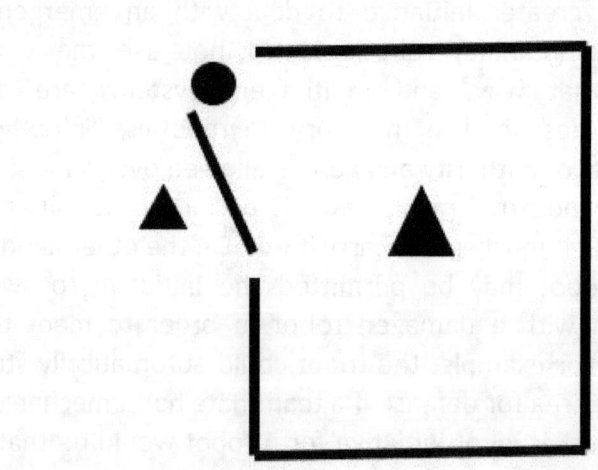

Figure 3: Frame from a Heider and Simmel Simulation. Although shapes move randomly, viewers see it as a story with the small shapes being chased out of the box (Heider and Simmel, 1944).

Returning to the idea of initiative, constraint relaxation requires the programmer to give the robot some initiative. However, that initiative is bounded and specific to the planning or problem solving algorithm. If the robot could not solve the problem by relaxing constraints, the program would return an error and do nothing. It would not react as Dave did and perform an unrelated activity like dancing to Gilbert and Sullivan.

So, if robots cannot spontaneously increase their level of initiative, why would it appear to humans as if they were showing more initiative? Because humans ascribe intent to any agent that moves. The psychologists Heider and Simmel started a series of experiments in the 1940s where they showed people a set of shapes randomly moving on a screen (see Figure 3). Even though the shapes were nondescript, non-humanoid, and moving at random, people would create entire stories around what they were observing. In many cases, they would explain that the large triangle was chasing the circle and the small triangle was trying to protect the circle or some similar story. Psychologists found that people could not help themselves: They attribute intent to anything that moves even when there is none. YouTube has several videos of variations on the Heider and Simmel simulation for viewers to see for themselves how hard it is to resist creating an explanation. In the case of "Catch That Rabbit," Powell and Donovan decided Dave was "taking initiative" and chose these words to attribute intention in order to explain the random behavior they witnessed.

Attribution theory is a subset of the larger phenomena of creating mental models of other agents. We previously discussed mental models in the chapter about "Reason." Attribution is essentially "mental model lite," where the mental model is very simple. We can easily sympathize and believe that Donovan immediately ascribed sinister intent to Dave when he observed what he thought were military drills. People make faulty assumptions all the time, and some are made with less evidence than Donovan's observations. It is interesting to note that even the best robot troubleshooters in the story jumped to unfavorable conclusions, even though they intellectually knew that the Three Laws bounded all the robots in their universe.

"Catch That Rabbit" illustrates the key principles discussed in the previous chapters, but it adds another HRI principle:

- **P6:** *Humans attribute intent to robots regardless of whether or not the robot is truly acting with intent.*

What humans observe influences their attribution of intent which, in turn, influences how they interact with the robot. As with the Heider and Simmel studies, people assume motion is purposeful and try to infer the purpose of that motion. The shape and other physical characteristics of a robot, the robot's morphology, impact this interpretation as we discussed in the previous chapter. A large black robot with small, red lights for eyes that walks directly toward you and stops after entering your personal space would be interpreted as if the robot were performing an aggressive action. A small white robot with large, blue eyes that performs the same movement would be interpreted as if the action meant the robot wanted attention, like a cute pet. In the same way, a robot that moves with jerky, start-stop motions is presumed to be dumber than a robot with smoother, more natural movements. Humans talk slower, using simple sentences, with a robot whose physical mannerisms match its internal intelligence. They also tend to get less frustrated with these robots.

Transparency and visibility should in theory help humans correctly attribute intent. In practice, they about algorithms and do not address how a regular person decides things such as whether or not a robot assistant is really there to assist and thus how far the person will trust the robot. Transparency and visibility are more likely to help robot designers and operators to trust their systems. But are robots trustworthy in general? That will be the topic of the next *I, Robot* story, "Liar!"

4.4 Questions

1. Define in your own words:
    - transparency
    - visibility
    - initiative
    - attribution theory

2. Was Dave a black box or white box?

3. How would you compare the ability Dave had to explain his processes with of the ability of Cutie (from "Reason") to do the same?

4. What is the difference between transparency and visibility?

5. Do you think the large number of movies and stories about robot uprisings encourages people to attribute sinister motivations to the actions of robots or do you think people can easily separate fiction from reality?

6. Do you think Sophia the robot is more intelligent than Atlas the robot? Why or why not? Can you explain your beliefs using attribution theory?

References and Suggested Further Reading

Heider, F., & Simmel, M. (1944). An experimental study of apparent behavior. *The American Journal of Psychology*, 57, 243–259. https://doi.org/10.2307/1416950

Chapter 5
Full Moral Agency: "Liar!"

"Liar!" has the distinction of being the first time the term "robotics" appeared as a word (Prucher, 2009). The story is about how RB-34, nicknamed Herbie, interacts with humans. While the previous stories have featured robots bounded by their task, Herbie is compelled by the First Law of Robotics to make everyone it meets happy. The conflict arises because Herbie can also read minds. Because of this, Herbie knows exactly what will make everyone happy. Unfortunately, as with all humans, sometimes the things we think will make us happy are difficult or impossible to actually achieve or obtain. Thus, to fulfill his definition of the First Law, Herbie tells lies that make people happy—at least while they believe the lies are true. It is a fun story and a great introduction to the ideas of ethics and morality in robotics.

5.1 Before You Read "Liar!"

Culturally, we do not think of robots as lying. Indeed, the media tends to portray robots as being painfully truthful. One of my favorite robots in fiction is Isaac from the television series *The Orville*. In "Into the Fold," Isaac bluntly explains to the two children trapped with him on a dangerous planet that they cannot go exploring with him because they are small, weak, and likely to die. This example with Isaac is a reminder that while we do not want or expect robots to lie as Herbie does in "Liar," we prefer them to phrase the truth in a more considerate manner.

Deliberate robot deception is one important area of research. In some cases, it is ethically and socially acceptable for humans to lie to each other, such as when someone asks about his or her appearance. Frequently, the truth is "stretched" or even ignored in cases like this so as not to offend the person asking for the information. For robots that do not have a real opinion about such questions but deduce answers from logic, a "lie" would be subjectively defined in cases of looks. However, questions such as, "Would it be acceptable for your healthcare robot to lie to you if the lie got you to lose weight?" are considered. In this case, if there is not a true, logical statement that would not produce the desired results, but there is a false statement that would produce those results. Is it ethically acceptable for robots to be programmed to give the false statement?

On the other side of robot deception, there is research by Cindy Bethel that shows that humans do not lie to robots. She has examined how this can be leveraged to discover the truth about crimes after the fact without children being afraid to tell the truth or being subject to human interviewer biases that may affect witness testimony (Bethel, Eakin, Anreddy, Stuart, & Carruth, Russell, 2013).

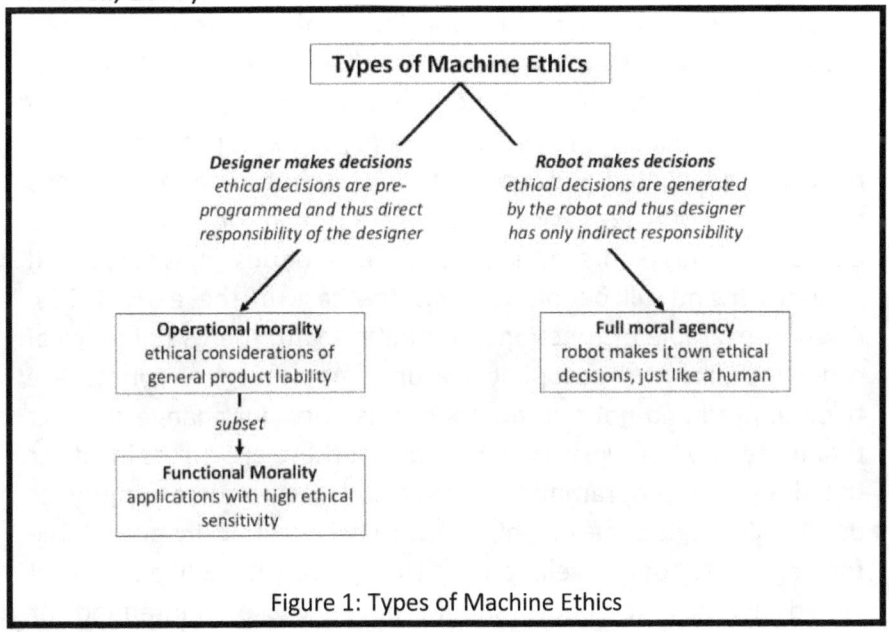

Figure 1: Types of Machine Ethics

According to Wallach and Allen (2008), ethical behavior, as related to robotics, falls into three categories: full moral agency, operational morality, and functional morality. These categories are explained pictorially in Figure 1.

The figure shows that in machine ethics, there is a distinction between whether we are discussing the ethics of the designer or whether we are discussing the preprogrammed ability of the robot to make its own ethical decisions. The right side of the diagram is about the ethics that a robot would use if it could make its own decisions; such robots are said to have full moral agency. These robots and the general artificial intelligence needed for programming them do not exist today and may only exist far in the future. Therefore, at this time, anything that any robot does involving ethical considerations is solely the responsibility of the designer, the left side of the diagram.

Designers are responsible for both the operational and functional morality of a robot. Operational morality is the ethical responsibility of a designer, as a professional, to design the robot to work correctly under reasonably foreseeable situations. Functional morality is the responsibility of the designer if the robot is being used for functions that have a high ethical sensitivity. An example of a task with high ethical sensitivity is the airplane autopilot feature. When engaging autopilot, a human pilot has delegated autonomy to the robot. The human has expectations of its behavior, and if the autopilot does not work as expected or has errors in the programming or design, it could lead to a crash and kill people, as was the case in the early 1970s. Another example of functional morality is with the use of surgical robots. Unlike the autopilot feature, most surgical robots are teleoperated and not autonomous. This does not change the fact that if they do not work as expected or if there are errors in either the design or programming, they could cause human injury or death. Although some might argue that it is ethically acceptable for certain robots developed by the military to kill people, all would agree that robots should never cause unintended or accidental human death.

5.2 As You Read "Liar!"

As you read the story, try to determine what type of moral agent Herbie is. Watch for the actions it takes and whether those actions have ethical sensitivity. Who takes responsibility for Herbie's actions? Also look for the role the Three Laws, which supposedly guarantee ethical behavior, play in Herbie's choices.

5.3 After You Read "Liar!"

Herbie was a full moral agent who made decisions for each local situation that did not work for the global situation. Many a person has made the same mistake but fortunately their termination is usually not an option. Of course, in Herbie's defense, he had only the Three Laws of Robotics to guide him. Humans have civil laws, religious laws, parental and societal norms, and lots of life experiences to guide us. So although Herbie was a full moral agent, the designers failed in their operational morality to consider the consequences of his rather incomplete moral upbringing.

Wallach and Allen (2008) proposed that if robots were full moral agents they would acquire a sense of ethics in one of four ways. These four ways are illustrated in Figure 2 and described below.

- **Consequentialism:** The robot reasons about the consequences of its actions and chooses the most ethical action based on this reasoning. In the story, Herbie seemed to apply consequentialism, but its view of the consequences was too limited. In real robotics, artificial intelligence is very good about reasoning and planning (consider chess-playing robots). The challenge is giving the robot sufficient knowledge about the world and representing that knowledge in a way that makes it easy for the robot to discover the correct solution using logical thought processes.

- **Deontology:** The robot acts according to its obligations, duties, and rights. Rules of ethics are established, and it basically follows these rules. Herbie also seemed to apply deontology, but although the Three Laws were ethical rules, they were also ambiguous and implicitly contradictory. Rule-based systems are famous for quickly becoming unmanageable and for having hidden contradictions.

- **Virtue ethics:** The robot develops a good character by observing others (who are hopefully good role models) and by learning from them. The robot observes how other saintly or highly respected agents (those people humans "put on a pedestal") act and then learns what these agents value and adopts these values for itself. This would require analogical reasoning where the robot would have to make an association between what it had observed and its current situation. Analogical reasoning is very hard to program into robots even though this is a common method of human learning.

- **Bottom-up:** The robot learns over time and through experience how to be good and make ethically correct decisions. Learning over time is hard. Machine learning has made significant progress in recognizing objects, but learning complex patterns that may change over time is challenging. Herbie may have learned a valuable lesson, but his bottom-up approach to learning ethical behavior was cut short. In general, humans do not expect robots to make ethically bad decisions in the name of learning even if this is also a primary method of human learning. Consider this: If a child throws a temper-tantrum in a store because it cannot have a toy, a human can pick the child up and carry it out of the store or even tell the child he can have a treat if he stops. If a robot throws a temper-tantrum in a store because it cannot have, say, a new memory chip, human methods of teaching and containing the tantrum would not work.

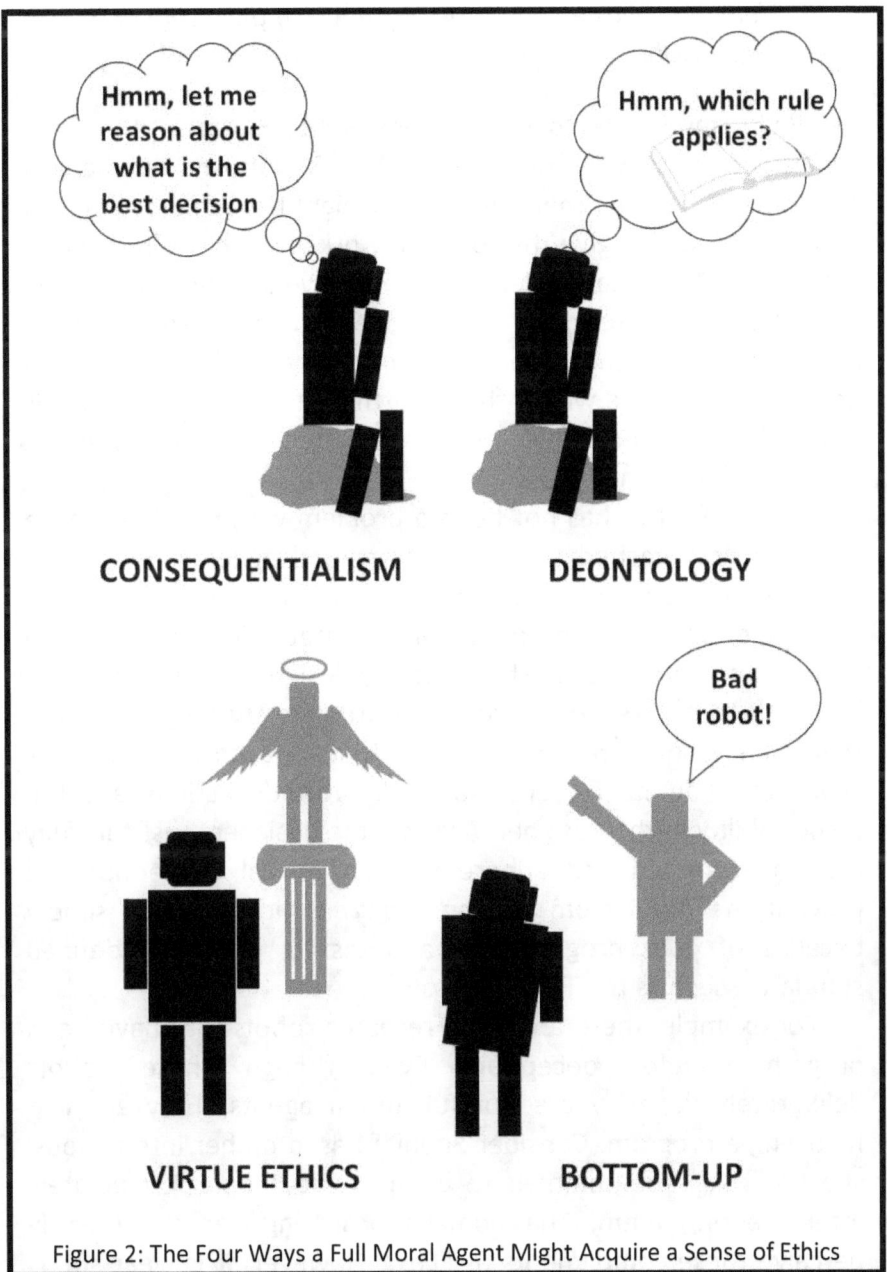

Figure 2: The Four Ways a Full Moral Agent Might Acquire a Sense of Ethics

From "Liar!" we can add another principle to our list:

- **P7:** *Every robot, even those not used for ethically-sensitive functions, has ethical aspects for which the designer is responsible.*

While Herbie's telepathic abilities were a surprise, the story makes it clear that the designers at U.S. Robots and Mechanical Men did not consider that their robots might have to act in ethical ways. They did not give the robots a workable ethical framework beyond the Three Laws. Herbie did not have sufficient reasoning capabilities to determine the long-term consequences of his actions. At the same time, the deontology of the Three Laws drove him to do the wrong thing. Further, he had no role models to develop virtue ethics, and he was too young to have developed ethics in a bottom-up fashion.

In real life, this has not been a problem with robots because we have not reached the level of artificial general intelligence required for some of these ethical issues to develop. Instead, modern ethics problems occur with operational and functional ethics. Designers often push out poorly designed or buggy robots in a rush to be first to market or in order to try to gain market share of a rapidly expanding market. Until AI becomes much more advanced than its current state, the robot's actions are the responsibility of the designer. As such, the designer must carefully test the program and adhere to high ethical conventions to prevent the robot from making a poor ethical decision simply because of poor programming and testing or poorly defined standards, such as the Three Laws.

For example, there are some research robots that have been programmed for deception. Even though these robots deliberately lie, they are not full moral agents. They are just following a program. Consider Sophia: Based on her interactions, she has been programmed to compliment people immediately after meeting them. The humans observing her could easily perceive these compliments as "lies." However, it is clear she is choosing from a list of preprogrammed compliments in her effort to meet this requirement of her programming. The difference between robots like Sophia and the robot, Herbie, in "Liar!" is that Sophia is not truly deciding to lie on its own, and the "lies" are socially acceptable.

The question of "Is it right for a robot to lie?" is really a question of "Is it right for a designer to program a robot to lie?" And this question leads to deeper ethical questions, such as "Is it right for a designer to program a robot to kill?" The next story, "Escape," shifts from such dark questions and provides a humorous look at user interfaces.

5.4 Questions:

1. Define:
   - full moral agency
   - operational morality
   - functional morality

2. Herbie is a full moral agent, but it was designed and built by humans. Under operational morality, are the designers responsible and liable for a full moral agent like Herbie?

3. In review of what we have already learned in "Catch That Rabbit!," did the troubleshooting for Herbie use white box or black box methods?

4. The robots in the *I, Robot* cycle are viewed as tools to get a job done even though they are often very intelligent, have self-esteem, and are even self-aware. Is this the same as treating robots as slaves? Are the people in the stories acting with good moral values toward robots?

5. If U.S. Robots and Mechanical Men, Inc., taught each robot it produced a system of ethics, would that have prevented Herbie from lying about hurtful things? Or would that have been a waste of time and resources because all the other

robots produced by the company were not full moral agents?

6. Can a robot be autonomous and not be a full moral agent? Explain your answer.

7. What virtues would a robot learn from Calvin or from Donovan and Powell if virtue ethics were programmed into the U.S. Robots and Mechanical Men robots and those people were defined as the pillars of virtue to be emulated?

8. Herbie learned the hard way that it should not lie, a bottom-up method of learning ethics. Clearly Calvin was in an irrational state of mind when she decided to decommission Herbie by presenting him with a conundrum. It is perhaps because of this that that she told this story in the context of "something going wrong on her watch." Without Calvin's interference, should U.S. Robots and Mechanical Men, Inc., have allowed Herbie to continue to learn or should they have decommissioned it? If they kept it in service, do you think it would have learned to avoid using its telepathy to understand humans and their commands or would it have continued to struggle with the Three Laws? Defend your position.

References and Suggested Further Reading

Bethel, C. L., Eakin, D., Anreddy, D., Stuart, J. K., & Carruth, D. (2013). Eyewitnesses are misled by human but not robot interviewers. In **2013 8th ACM/IEEE** International Conference on Human-Robot Interaction (HRI), pp 25–32. IEEE.

Prucher, J. (2009, March 31). Nine words you might think came from science but which are really from science fiction [Web log post]. OUPblog. Oxford University Press. Retrieved December 14, 2019 from https://blog.oup.com/2009/03/science-fiction/

Revell, T., (2017). Robots could help children give evidence in child abuse cases. New Scientist. Retrieved Dec. 14, 2019 from https://www.newscientist.com/article/mg23331184-500-robots-could-help-children-give-evidence-in-child-abuse-cases/

Swanwick, M. (2016). The scarecrow's boy. In Not so much, said the cat (pp. 43-51). San Francisco: Tachyon.

Wallach, W,, & Colin A. (2008). Moral machines: Teaching robots right from wrong. New York: Oxford University Press.

Chapter 6
How Robots Are Programmed: "Little Lost Robot"

Although the title "Little Lost Robot" sets up an expectation of the need to rescue a cute, helpless, lost robot, this story is really about a very capable robot, Nestor-10, who is angrily told to get lost. The programmers modified the First Law to eliminate the part about a robot having to prevent a human from being harmed. The modification was made because the original Nestor robots kept "saving" humans from non-lethal exposure to radiation during radiological experimentation. The situation is similar to a robot rushing into a building, seizing a firefighter, and dragging him or her away from a fire. It would be counter-productive to say the least! However, this minor change to the Three Laws had unintended consequences and allowed Nestor-10 to become resentful of its human masters. When told to lose itself, it did so and then worked hard to stay hidden by pretending to be one of 62 other unmodified robots that were cargo on a visiting ship. In the story, U.S. Robots and Mechanical Men, Inc., send in Peter Bogert and Calvin to find Nestor-10. Bogert, as the engineer, only knows about programming frameworks suitable for verifying the robot works, but Calvin, as the robot psychologist, knows both about programming frameworks and about how to predict the resulting human-robot interactions that could occur as a result of that programming. Thus, "Little Lost Robot" is a great opportunity to explore how real robots are programmed and how that might impact human-robot interactions.

6.1 Before You Read "Little Lost Robot"

In real life, AI-based robots are not programmed with rule-based systems where everything is controlled by if-then rules. Rules such as "If robot senses S, then do A," usually do not work. One reason is that rules use symbols. For example, if CAR_ROBOT senses PEDESTRIAN, then do SWERVE. (It is convention in AI to capitalize symbols.) But a robot typically senses a pattern with a probability that it is a pedestrian instead of a binary it-is-or-it-isn't response. So, at what value of certainly should the car swerve? 100%? 2%? One approach is to make the rules into fuzzy rules and use fuzzy logic to decide.

Another problem with rule-based systems, fuzzy or not, is that the world and the spectrum of actions the robot should perform are too complex to represent with predetermined rules. In addition, humans generally cannot anticipate what rules are needed to cover all the possible cases and how to correctly state the rules. Even if all the rules are correct, the order in which they are stored in the rule base effects the outcome. For example, the robot might first encounter a rule that fits the situation but is very general. Since this rule is listed first, the robot will obey it and never consider a rule that is more specific. For example, it might find if CAR_ROBOT senses PEDESTRIAN, then do SWERVE before finding if (CAR_ROBOT senses PEDESTRIAN) and (CAR_ROBOT senses CAR_IN_NEXT_LANE), then do HARD_BRAKE.

Instead, artificial intelligence researchers often organize a robot's programming framework into three layers, enabling a robot to concurrently perform three different aspects of intelligence. This arrangement is similar to how we believe intelligence is organized in the brain and central nervous system of humans. Figure 1 shows a sweeping generalization of how intelligence is organized and how that translates into an operational architecture. The programming framework is referred to as a software operational architecture to distinguish it as the generic template for a specific model of robot, and not the actual detailed technical architecture. This is similar to how all the U.S. Robot and Mechanical Men robots are essentially programmed the same, but with different models, such as Nestor, getting different modules for their specific application.

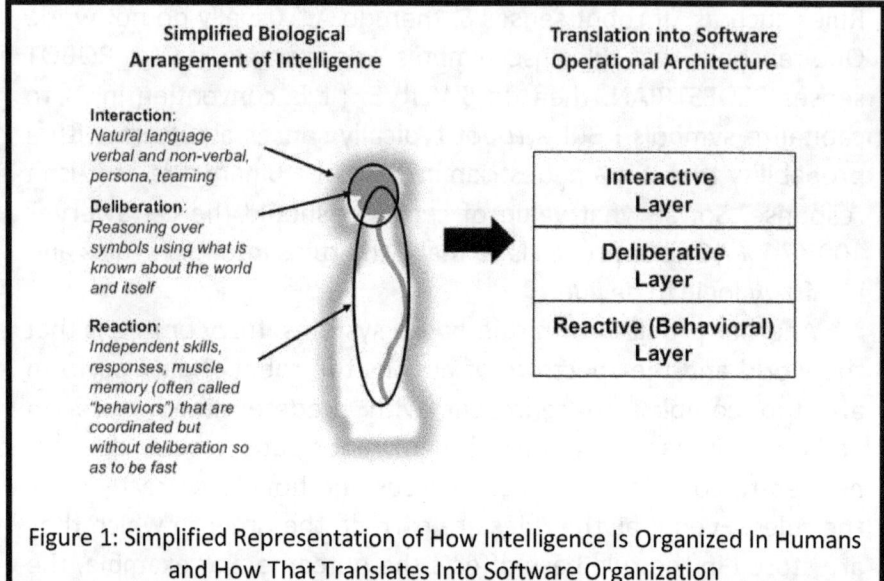

Figure 1: Simplified Representation of How Intelligence Is Organized In Humans and How That Translates Into Software Organization

The lowest layer is "reaction." You could compare it to the part of the central nervous system that produces reflexes. These rapid sense-and-then-act loops are generally called "behaviors." Lower order animals and single-celled organisms use only reaction to process and analyze information about the world around them and are successful. Robots work with only reaction. For example, the early versions of the iRobot Roomba™ merely reacted to their environment. This allowed them to vacuum rooms they had never seen. The Roomba™ was not optimal in that it did not clean a room in the shortest time possible or with the least amount of energy, but it did get the job done without human supervision.

"Deliberation" refers to the cognitive abilities we associate with the human brain. Deliberation typically relies on some type of internal representation or world model, as previously discussed. The internal representation may be propositions or facts about the world. It is important to note that these facts use symbols or labels instead of the raw sensor inputs. For example, my coffee cup might be represented with CUP(location, ROBIN), not the sensor values of the image system when focused on the cup. Converting these sensor readings into symbols is still a major barrier in computer vision. In "Little Lost Robot" when Nestor-10 was being trained, the engineers would have uploaded the different sensor readings for gamma radiation and infrared radiation, but these would have been stored as symbols, perhaps GAMMA and INFRARED.

Once these are in place, a reasoning mechanism can search the set of propositions to make inferences, solve problems, or generate new goals. Significant research has gone into making sure the resulting inferences are provably correct, that is, that the reasoning mechanism is working correctly. Unfortunately, it has been much harder to make sure the propositions are correct (otherwise you get effects like those in "Reason"), to change or to adapt propositions over time as things change in the environment that affect the world model (e.g., a store closes, the family pet dies), and to handle the increasing number of propositions. Deliberation often produces a plan that the reaction loop carries out.

"Interaction" is the social wrapper around the agent, capturing what the robot interprets about other agents (relatives, co-workers, angry barking dogs, etc.), how it communicates with these agents, and teaming. As we've discussed in "Robbie," people use both verbal and non-verbal forms of expression. Robots may use one or both forms. Those methods of communicating may be bundled into a role that exhibits a particular demeanor. For example, a robot butler would likely act in a differential way than a robot fitness counselor. This aspect is called a "persona." Teaming can be anything from a hard-wired mating in insects to a person covering for a sick co-worker. Insect mating does not require deliberation because it is hard-wired into the insect like reflexes. Taking over for a co-worker would require deliberation because it would entail you examining what needs to be done (as opposed to those things that can or have to wait for the co-workers return) in terms of the work you are performing for him or her and balancing that with your own work. Interaction between humans and robots always requires deliberation, and can involve reasoning, beliefs, desires, and intents. It also requires reactions to manifest teaming and non-verbal expressions.

Notice that each layer of intelligence can see and modify the layers below it. For example, a robot that helps an elderly person might have a cheerful interaction and to maintain that cheerful disposition would require internal deliberation and reaction. Nestor-10 interacts with Bogert and Calvin to understand what they intend, then deliberates by sorting through the facts about his programming in order to create a plan that does not betray his identity. The plan is executed by Nestor's behaviors.

One subplot of "Little Lost Robot" is the issue of superiority. Many humans in the Western world fear that robots will become superior and attempt to usurp human "masters." This fear has been present in many of Asimov's stories we have covered. In "Little Lost Robot," however, you will encounter humans referring to the robots as "boy." Written in 1947 (prior to the Civil Rights Movement in the United States), this term was meant to evoke the feeling of a slave/master relationship and was used by whites, especially in the South, when addressing an African-American man of any age. Although its use and the history of it may be shocking to us now, it was a reasonable reflection of society at the time that robots would be likewise treated by many people as an "inferior" race.

6.2 As You Read "Little Lost Robot"

As you read the story, watch for the discussions of how the variants of the Nestor robots are programmed. What do Bogert and Calvin know about the differences between the original Nestor robots and Nestor-10? What other differences are pointed out throughout the story? Note how Bogert and Calvin used their understanding of the robots' programming to try to identify the lost robot. Also think about the ways you would have dealt with the problem of the robots sacrificing themselves trying to save people from non-lethal levels of radiation. Was there a better alternative than the one used in the story, which was to modify the First Law.

6.3 After You Read "Little Lost Robot"

The story was a contest between mathematical and psychological understanding of human-robot interaction, with Calvin as the winner. Bogert concentrated on the mathematical correctness of Nestor-10's programming. While it was good to know that it was provably correct with automated reasoning that Nestor-10 would not harm a human, that result was not particularly interesting or helpful. It was more of the same design philosophy used by the U.S. Robots and Mechanical Men, Inc., engineers when they first implemented the modifications. Bogert was unable to think "outside the box" and sometimes that can prevent discovering the answer to a problem. Calvin took a robopsychological approach. In real life, psychological approaches in HRI are usually human-centered; they focus on designing a robot so that a human is comfortable with how the robot accomplishes its goals. Calvin, as a fictional robopsychologist, was robot-centered and used her knowledge of how robots were programmed to set up the trap for Nestor-10.

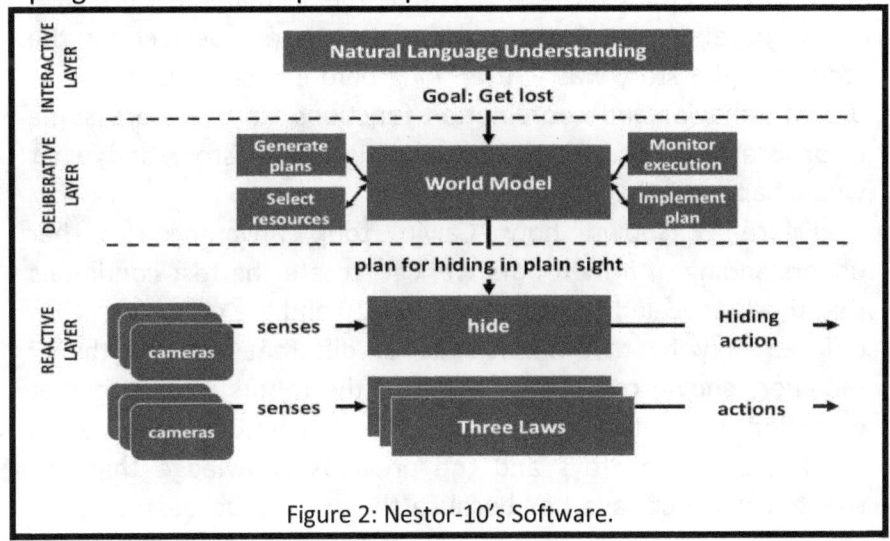

Figure 2: Nestor-10's Software.

Figure 2 shows a more detailed software operational architecture describing the programming of Nestor-10. It was strongly told verbally and non-verbally to get lost, which its interactive layer processed and passed to the deliberative layer as a goal. The deliberative layer used its understanding of the world (stored as a world model) to generate a plan for hiding, selecting the best behaviors to carry out the plan, implementing those behaviors by telling the reactive layer what sensors and behaviors to activate, and then monitoring the plan and revising the plan when Nestor-10 needed a different hiding strategy.

Behaviors for the Three Laws were always active in the reactive layer. Note that these behaviors are concurrent, that is, they are operating in parallel. This means that a robot could be smart enough to "walk and chew gum at the same time." Interactions between behaviors are governed by a coordination function. This example uses the subsumption architecture where behaviors can subsume and inhibit other behaviors. (We can read the story as Bogert focusing on proving the coordination function was mathematically correct, not what the actual outcomes of the coordination would be.) The Three Laws would subsume any behaviors to save a human or itself, and the Three Laws would subsume each other, with the First and Second Law overriding the Third Law and the First Law overriding the Second and Third Law. Please note: The choice of subsumption is not perfect for the story, but the story was written long before roboticists began to design behavior and coordination functions, so there are some inconsistencies between what happens in the story and what would happen if the story were true.

Figure 3 shows how Calvin took advantage of her understanding of how robots work to create the test conditions that finally revealed Nestor-10. Nestor-10 and the other 62 robots differed only by their knowledge of different wavelengths of radiation, shown crosshatched. When the robots were told that gamma rays would be between her and them, the information about the gamma rays and the previous knowledge that the robots could not save her because they would be destroyed by the radiation set up the software relationship shown in Figure 3. The interactive layer passed on the information that any radiation would be gamma rays. The deliberative layer implemented the Third Law behavior with this information as a parameter, something that added to the behavior but did not change its fundamental operation. Now the Third Law could inhibit (symbol "I") the First Law. If a regular robot sensed any radiation at all, it would assume that radiation was gamma radiation. If it sensed the radiation and at the same time sensed danger to a human, it would not destroy itself by entering the radiation in a futile attempt to save the human.

The key was the information the robots had. Although all the robots could detect radiation, they did not all have the knowledge to associate different wavelengths of radiation with different kinds of radiation. On the other hand, Nestor-10 had been trained to distinguish between gamma rays and other forms of radiation, while the others had not received this training. Therefore, the information from the interactive layer that if the detector detected any radiation, it must be gamma rays was extra information, not a replacement for the robot's radiation sensor.

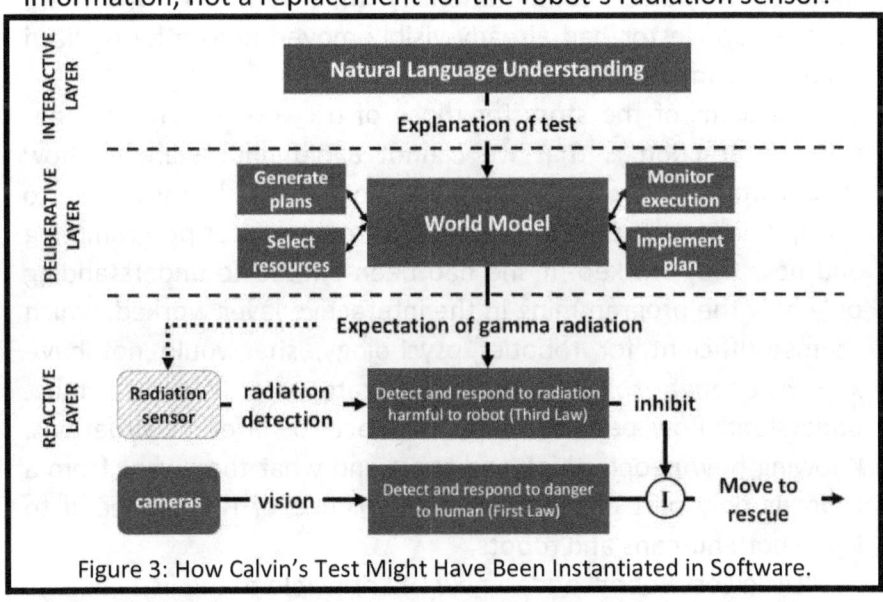

Figure 3: How Calvin's Test Might Have Been Instantiated in Software.

What Calvin did would be like telling a runner not to trust his or her eyes or learned skills and to always go right to avoid puddles. Since the skills of a runner are based on perceiving the best footfalls reactively, and not explicitly deliberating about each step, the human runner would probably ignore the advice and run reactively as he or she normally would. However, once it got dark and the runner had to mentally try to deliberate about what he or she was seeing to make up for his or her impaired vision, the runner might take the given advice and go to the right for each puddle. But a human runner would always trust his or her senses and training over advice.

Likewise, Nestor-10 used its training in combination with its sensors and not the information given to it by people trying to find it, and thus moved reactively to save Calvin. The deliberation layer was also monitoring the reactive layer to make sure the plan of staying hidden was working, however. As a result, the robot quickly stopped moving when it realized none of the other robots had reacted. It first tried to resume hiding and then tried a new goal of attacking Calvin. But deliberation, which is conscious thought, is slower computationally than reaction, which are reflexes, so Nestor had already visibly moved before he realized his error. Game over for Nestor.

The point of the story for those of us working with human-robot interaction is that we cannot avoid understanding how robots work. We can infer from the story that Calvin was able to set up this test by understanding all three layers of programming and how they worked. If she had been limited to understanding only how the programming in the interactive layer worked, which seems sufficient for robotics psychology, she would not have known enough to set up an accurate test. We also must understand how people think and react to different situations. Knowing how people think and react and what they want from a robot is only part of meeting the challenge of HRI, we need to know both humans and robots.

"Little Lost Robot" adds a new HRI principle to the list:

- **P8**: *It is difficult to create or debug an effective HRI scheme without understanding how robots work.*

The First Law was a fundamental principle of the HRI scheme, yet it was changed without understanding the subtle impact of a fairly common or predictable interaction—that a worker would get mad at a co-worker and say something rude. The situation was resolved only because the non-engineer, Calvin, understood enough of robotics to create a trap for the lost robot to reveal itself. Likewise, human-robot interaction practitioners coming from a psychology or communications background may need to become more familiar with how robots are programmed in order to be successful.

Certainly a person cannot be an expert in all aspects of human behavior plus all aspects of robotics, but focusing on understanding just one, either humans or robots, is not sufficient. Most of the *I, Robot* stories illustrate how engineering-oriented designers forget the human aspects of their work while "Little Lost Robot" illustrated the dangers of forgetting the robot. The next story, "Escape," provides an opportunity to discuss how a designer might create a workable HRI scheme combining both psychology and engineering.

6.4 Questions

1. Describe the difference between Bogert's and Calvin's approaches to troubleshooting.

2. Should Calvin have predicted Nestor-10 would attack her? Why or why not?

3. Is the fact that the humans found the modified Nestor robots annoying a sign of a flawed human-robot interaction? Was the fact that humans were annoyed predictable?

4. Could you represent the information that any radiation would be gamma rays as a behavior that subsumed the Third Law, rather than modified that? Would they create a problem by being able to override the Three Laws versus just modify them with parameters?

5. How does this story examine transparency in troubleshooting? How could a designer have used visibility to recognize the modifications were going to cause problems in future interactions?

6. Did verbal or non-verbal communication or both help to reveal which robot was Nestor-10?

7. In the story, humans viewed themselves as superior to robots, but some of the robots considered themselves superior to humans. Based on the other information revealed, do you feel the humans or the robots were superior, or were they equal?

Chapter 7
User Interfaces: "Escape!"

Can you image getting onto an airplane with no windows, no displays, no flight attendants or any indication another single living person is on the plane with you, no indicator of how long you are going to be traveling or the destination to where you are going to travel, not even a door for a restroom—nothing but a cold metal tube and one chair in the middle for you? "Escape!" has two familiar characters, Donovan and Powell, experiencing a trip, not in an airplane, but in a spaceship as bare as the airplane just described. Needless to say, this sterile interaction is a nerve-wracking experience as Donovan and Powell find out in "Escape!" But the key is that the spaceship and the human-robot interaction scheme were designed and controlled by an artificial intelligence system called "Brain." Despite being more intelligent that humans, Brain made the same mistakes human designers do: It ignored the human-robot interaction scheme. "Escape!" is a marvelous introduction to the user interface portion of a HRI scheme.

7.1 Before You Read "Escape!"

The term "user interface" (UI) is often used synonymously with "graphical user interface" or GUI. A GUI is a screen that displays data from the robot and allows a user to interact with it. However, there are multiple other modalities of user interfaces. The robots in Asimov's world all tend to directly interact with their users without the use of a screen. For this reason, we are not surprised that Brain is lacking a GUI when Calvin approaches it to interact. Instead, she directly speaks to Brain using a naturalistic interface. Naturalistic interfaces are much less common outside of science fiction because of the difficulty of developing verbal and non-verbal communication mechanisms. On the other hand, we would expect even the futuristic spaceships of Asimov's world to contain some sort of physical interface, such as buttons or a joystick or a combination of a GUI and physical interface so those riding inside it could assume control if needed. A third type of user interface is a brain-computer interface (BCI). This is something that evokes the memory of Luke Skywalker getting a robotic new hand, but it has actually become a reality that is at the forefront of prosthetic limb research. Currently, these are only available to amputees as methods of controlling their prosthetic limbs inside a research study, though. They have not yet been cleared for distribution to the general public.

An interface has three difference purposes regardless of its modality and this means a designer may have to build three different interfaces. The most obvious purpose is "end-user execution"—that is to enable the end-user to operate the robot. In the case of the Brain, its naturalistic interface allows Calvin to give it a command to solve the problem of hyperspace travel. Of course, nothing goes as expected with Asimov's robots and this leads to another purpose for an interface: troubleshooting.

Troubleshooters (and developers) need specialized types of information and ways to engage the robot than what an end-user requires. Thus, there is a "developer diagnostic" interface that is often hidden from the end-user but can be activated by an manufacturer's representative as needed. If Brain provided a developer diagnostic interface, Donovan and Powell couldn't find it!

The third purpose of an interface is to explain or demonstrate what the robot is doing to the public, particularly the press and its sponsors. These "public explication" interfaces show what is going on with the robot that makes it special. For example, a public explication interface for a snake robot might show how a new algorithm computed and ranked all the possible shapes the snake could take in order to get through a narrow passage. A public explication interface would have at least given Donovan and Powell confidence that the ship was working as intended and that they were traveling to a set point and then returning to earth.

In practice, robot designers build a developer's diagnostic interface because they need it to test the robot, and then they assume that the interface can be reused as satisfactory end-user and public explication interfaces. End-users do not perform well with a diagnostic interface because these interfaces contain extra information that clutters the display and interferes with end-user cognitive processes. A famous example is "Dark Spot," the nickname for an expensive military prototype unmanned aerial system named Dark Star that crashed. The autonomous control software had encountered a glitch on take-off, but fortunately a human pilot stood ready to take control. Unfortunately, the pilot was stymied by the graphical user interface that had been designed for a developer. The GUI was presenting too much detailed data that was in the wrong form for the rapid cognition and decision-making needed to fly the plane. The pilot could not intercede in time to prevent the crash.

Figure 1a shows an example of the set of three types of user interfaces which were developed for Japanese search and rescue dogs. As shown in Figure 1a, each dog carries a backpack with a camera and biosensor to let searchers see what the dog is seeing and to alert the operator when the dog is getting tired. Figure 1b shows the end-user operation interface. Everything on the GUI is associated with data needed for conducting canine search and rescue; the left window shows where Dog 1 has already searched, while the right window shows what the camera is seeing and lets the operator take pictures. Since the dogs are all fresh and alert, none of the tabs for Dog 1, Dog 2, and Dog 3 are shaded with an alert.

Figure 1: (a.) A Cyber-canine System with Cameras and Biometric Sensors in the Orange Pack; (b.) the resulting end-user execution GUI that a search team leader sees; (c.) the public explicative interface shown to sponsors and the press, and (d.) the developer diagnostic display (Murphy & Tadokoro, 2019)

Although this interface is useful for the responders, a sponsor or a reporter cannot visualize what is special about the backpack since cameras have been put on dogs for over a decade. The backpack has two special features: It computes the fatigue level of the dog plus classifies what the dog is doing, such as sniffing or barking, which combined with the video data would be used to indicate areas of interest for human searchers. Thus a public explicative display was created and is shown in Figure 1c. Note that the public explicative display has some of the information from the end-user operation display but removes the controls and adds visualizations of the biometric status of the dog as moving graphs. In contrast, the developers prefer a more stripped down GUI that lets them work with the algorithms directly (Figure 1d).

7.2 As You Read "Escape!"

As you read the story, watch for descriptions of what Donovan and Powell expected in a user interface and what Brain actually gave them. In particular, think about what information they wanted in terms of an end-user operation and developer diagnostic interface. How would that information have been best captured by what combination of physical, graphical, and naturalistic interfaces? Another aspect of the story is that Donovan and Powell are inside the ship (robot) as opposed to Calvin who is dealing with a robot that next to her. Note the difference between the expectations of an interaction between the robot and human as agents external to each other and the expectations of an interaction with a ship (or self-driving car) where the human is contained within the robot.

7.3 After You Read "Escape!"

Wow, that space trip sounded unpleasant! On the other hand, it was amusing to see that intelligent machines might be as bad, or worse, at the human-robot interaction aspect of designing a spaceship. The Brain did not provide an end-user execution interface because there was no end-user per se, just passengers. But the passengers were not given any sort of public explication interface and there was no developer diagnostic interface for Donovan and Powell. Brain did not feel it needed to communicate with humans onboard the ship through natural language or any other means, and there were not even physical interfaces--- no knobs, dials, or joysticks— for Donovan and Powell to use during troubleshooting. On the other hand, giving the humans the same meal of milk and beans for breakfast, lunch, and dinner and providing bathrooms and access to the food only during certain times as determined by the ship was a form of communication: It was clear to Donovan and Powell that the humans aboard the ship did not matter.

Figure 2 shows an ideal design process that would help an HRI designer avoid creating an interaction scheme that would cause humans to be frustrated with robots. The HRI designer would begin by imagining the types of robots, what they would be used for, the kind of environment in which they will be working, how the humans would expect to communicate with the robots, and the capabilities of each agent. For example, a human passenger, regardless of whether they were robot specialists such as Donovan and Powell or a regular passenger who is just being transported, would want information about the flight, not blank walls. They also would expect some human control over their experience, for example, Donovan and Powell could only use the restroom when Brain decided to allow them that luxury.

In addition to not having these basic features, the ship was lacking in other interactive features that would have made human passengers feel at ease. For example, it is common for airlines to provide humans with an entertainment system that has an optional display showing where the plane is on a map, the distance traveled, and the time remaining in the flight. Many airlines realize that some passengers are pilots or interested in aviation and allow passengers to listen in on the air traffic control communications. Passengers are alerted a few minutes in advance that a meal will be served. Not only does this allow passengers to prepare for the meal but also subconsciously builds trust because the airplane then does what it says it will do.

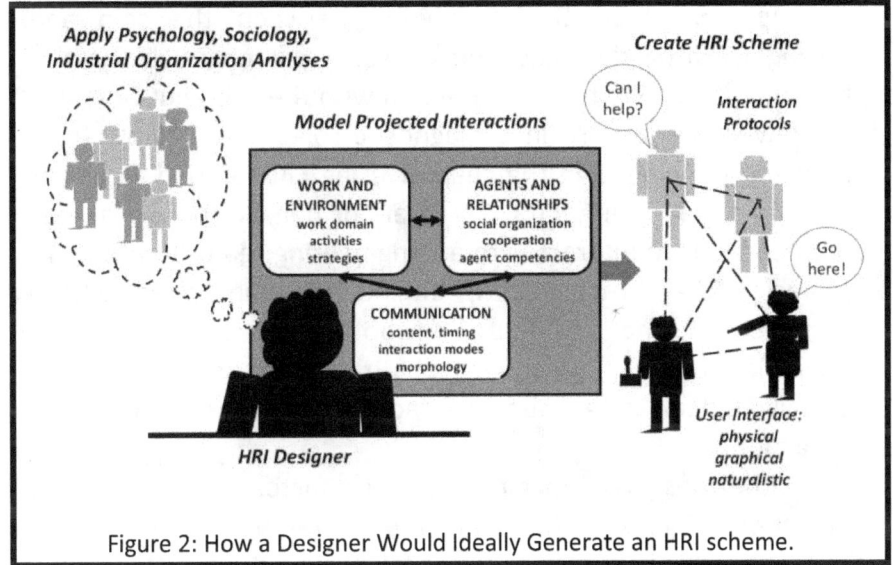

Figure 2: How a Designer Would Ideally Generate an HRI scheme.

Even in a hospital, one of the most austere environments for a human, most patients get a TV with a remote control, a clock, and a board with the current date and status of the patient. All these things help the patient feel more in control of his or her surroundings and provide a method of distracting them from the fact they are laying in bed, sick. Both passengers and patients are permitted to ask questions and expect to get answers.

Computer scientists learned decades ago that people need an indicator that a computer is working on a problem (the hourglass or spinning rainbow wheel) and a time to completion (the moving bar for a file upload). Without these indicators, humans jump to the conclusion that something is wrong, just as Donovan and Powell did in "Escape!" Brain provided none of this basic level of transparency and visibility expected by humans, which heightened the tension between the humans and the robot.

An HRI designer should also consider the roles of the humans throughout the entire lifecycle of their interaction with the robot, from how the humans and robot spaceship would interact when the mission was initiated to how it would be executed, what would happen if something went wrong or was suspected of going wrong and how the mission would terminate. The previous paragraph discussed the interaction between the ship and passenger during the execution phase. A passenger during this phase would have minimal interaction with the ship, but Donovan and Powell were there in a diagnostic role. As we know from discussing Principle 3 in the chapter using "Runaround," a robot always has the potential for a problem or bug, so there is always the need for it to interact with an engineering team (Principle 4). As seen in "Catch That Rabbit!," engineering teams need to have specialized information about the robot (recall transparency and visibility) and even direct control. Brain had no backdoor or normally hidden diagnostic interface to facilitate a diagnostic interaction.

As the designer imagines all the factors that influence interactions, Figure 2 shows a template for how those factors should be organized to help the designer produce a reasonable HRI scheme. The HRI scheme would need to be tested and refined, but the process should help the designer avoid the goofs seen in Asimov's stories. The designer would make formal observations and apply knowledge about the humans and robots in order to form a model of the expected interactions and any constraints to those interactions. The model should contain what the designer understands about the work to be performed and the environment (also called the work envelope), the agents and their relationships, and the forms of communication that they will use. The work and environment can be further divided into descriptions of the work domain, activities, and strategies.

To use an example to illustrate, consider again the Roomba™ robot that was created to vacuum houses when the owner is away. The robot and owner do not have any social organization or real cooperation and both are competent (the robot knows how to vacuum, the owner knows which rooms need to be vacuumed). The content of communication is minimal as the owner must turn it on, let it work, and then it turns itself off or returns to the charging station. The owner needs to be alerted to problems like when it is out of battery power or stuck. The timing is limited to the initiation, termination, and exception handling phases. The owner might like to use a natural language command system as the communication mode but generally prefers a low cost robot more than he or she wants the novelty of speaking commands to it. Having a robot competent in natural language adds expense, so designers chose a standard user interface. Fortunately the robot's appearance or morphology gives humans no signal that it is intelligent and will be able to do anything more than vacuum. This means a person will not expect a more social interface. As a result of the interplay between all these factors, the end-user interaction protocol relies on a simple interface: an on-off button, buttons to select the size of the room, a status light to know the robot is charged and working, and a sound to alert people that it is stuck. There is no need for a public explication display. A Roomba™ returned for repair has a diagnostic interface that the technicians can access.

This example is very specific in that it uses a robot that requires minimal human interaction. However, a social robot like Sophia or a teleoperated bomb squad robot would have a very different HRI scheme. The important thing to remember is that each HRI scheme must take into account all the different aspects of the specific robot in question. A scheme created for a Roomba™ would not work for an autonomous car where the user must be able to take direct control in emergency situations.

As we see, "Escape!" adds two additional HRI principles to the list:

- **P9**: *A user interface can be either for end-user execution, for a developer's diagnostics, or for explication to the public.*

- **P10**: *A human-robot interaction scheme is based on understanding the interactions between work and environment, agents and relationships, and appropriate communication mechanisms.*

These principles explain the flaws in Brain's spaceship design. Principle 9 reveals that although Brain did not need to provide an interface for the execution phase, it did need interfaces for the public (passengers) and the developers (Donovan and Powell). In most cases, this reflects the current practice in robot design where designers try to retrofit the human-robot interaction scheme after the robot is constructed. The HRI is then based on feedback from the users. Principle 10 discourages building a robot and then discovering that the HRI scheme does not work. Iteratively developing a good interaction scheme can be time consuming, and it can be expensive if the suitable HRI scheme requires the manufacturer to add more sensors or redo software. It is possible to project what different types of interfaces humans will need for a task and how they will want to interact with the robot so that the resulting scheme requires only minor modifications.

"Escape!" was about a robot providing no user interface combined with a hostile morphology. The Roomba™ example briefly touched on morphology as one form of communication.

The next story in the *I, Robot* collection introduces a district attorney, Stephen Byerley, who may or may not be a humanoid robot. If Byerley is a robot, its successful masquerade is based on how well its intelligence and other communication mechanisms align with what humans expect of other humans.

7.4 Questions:

1. What is the difference between an end-user execution, developer diagnostic, and public explication interface? How would have you designed each of these interfaces for the Brain's ship?

2. Do you think a system such as a Mars rover would have different interfaces, and if so, which ones would be necessary?

3. How is an autonomous space ship similar to an autonomous car? What is the relationship between a driver and a self-driving car? What can each agent (driver, car) do better than the other?

4. What kind of user interface does a non-autonomous car have? How do you communicate with the car? Is there a difference between communicating for control decisions (stop, turn left, change speed, etc.) and communicating for non-driving activities (reading text messages, playing music, etc.)?

5. Consider what you know about driving a car. How would you characterize the work and environment for driving a car along an interstate highway? How would the work and environment be different for driving a car in the middle of a busy city?

6. How would the interfaces for a health care robot that could pick up an elderly person up out of a bed be different from the interfaces for a robot pet? What do you know about elderly people that would influence the design of a health care companion? For example, do they have hearing problems that would make it hard for them to hear a robot that only communicated verbally?

7. Was the type of information that Donovan and Powell wanted from Brain something that a designer could have predicted that a passenger would want? Is there additional information that you, as a passenger, would have wanted?

8. Is it ethical to allow an artificial general intelligence that does not understand humans to design a system for humans?

9. Consider that sometimes troubleshooting is done in teams. Although Donovan and Powell always seem to work

together, Calvin wanted to do the troubleshooting with Brain alone. Discuss the potential benefits and disadvantages of working as a team and working alone to troubleshoot a HRI problem.

References and Suggested Further Reading

Murphy, R.R., & Tadokoro, S. (2019). User interfaces for human-robot interaction in field robotics. In S. Tadokoro (Ed.), Disaster robotics: Results from the ImPACT tough robotics challenge (pp. 507-528). Switzerland: Springer International Publishing.

Chapter 8
The Uncanny Valley: "Evidence"

"Evidence" is a clever story that asks how could you prove a really nice, moral person was not a robot. A local district attorney, Stephen Byerley, is running for mayor after recovering from a major car accident. His opponent starts a smear campaign, claiming that he must be a robot because no one has ever seen him sleep or eat. Short of dissecting him, a clear violation of a person's civil rights, how can either side prove that he is human or a robot? "Evidence" serves as a great introduction to the Uncanny Valley.

8.1 Before You Read "Evidence"

If Byerley is a robot, he is indistinguishable from a human and has avoided the Uncanny Valley. The Uncanny Valley a term used in human-robot interaction to explain why robots sometimes appear to be creepy or unsettling. Masahiro Mori first proposed the phenomenon in 1970 and explained it with the graph that has been simplified in Figure 1. "Human likeness" is along the x-axis and "Familiarity" is located along the y-axis. In this case, "Human likeness" refers to how closely the morphology of the robot resembles a human, where a robot like the Roomba™ would score low and a robot like Sophia would score high. "Familiarity" means how comfortable a human is with the robot and the scale ranges from "finds it creepy" to "totally comfortable with it."

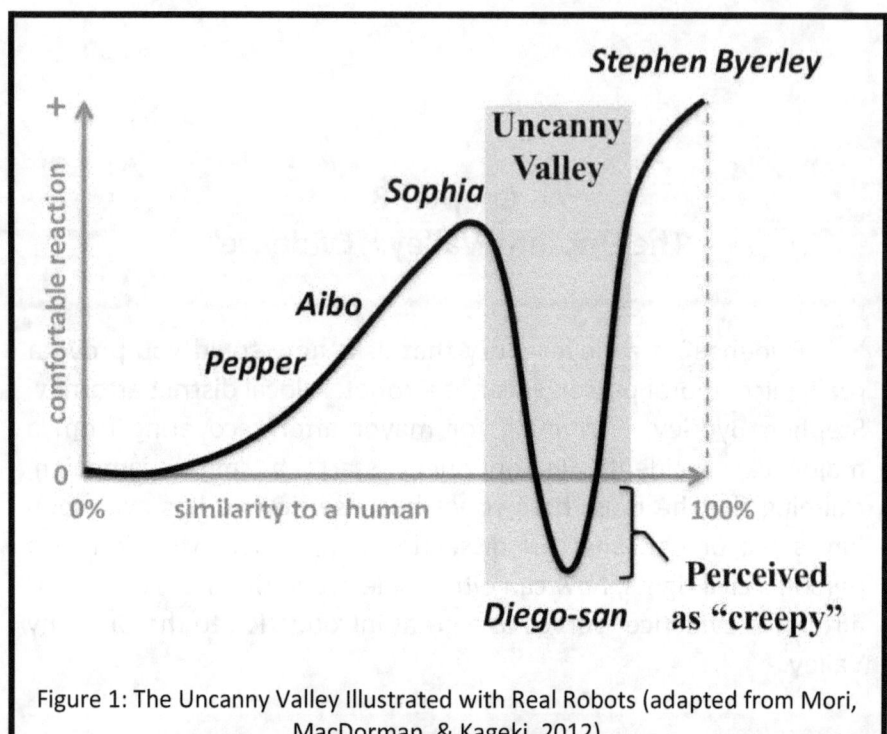

Figure 1: The Uncanny Valley Illustrated with Real Robots (adapted from Mori, MacDorman, & Kageki, 2012)

Mori observed that animated toys, puppets, robots, and computer animations all follow the same pattern of eliciting emotional reactions to their appearance. Puppets do not look exactly like animals or people, so when they do not move exactly like animals or people that is acceptable because their appearance signals that they are only approximations of animals and people. If a robot looks like a human but has a slightly bigger head, like the Diego-san robot (Figure 2), which was built to replicate a small child but whose head is not proportionally the right size (it is bigger than an adult's head), then most humans will classify the robot as "creepy."

Creepiness is not limited to just how well a robot reproduces the look of an animal or human, it applies to its expressions of intelligence. If the robot is a humanoid and is fairly primitive in terms of how much it looks and moves like a human, then the robot will not be considered too creepy. This is because the appearance matches the actual, relatively low, intelligence of the robot. In entertainment media, humanoid robots are often portrayed as moving slowly, more deliberately, or jerkily. This signals that the agent is a robot and will have some sort of limitation. Robots that look identical to a human but have some sort of unnatural capability, like twisting their head 360 degrees, are fascinating but rank high on the creepy scale.

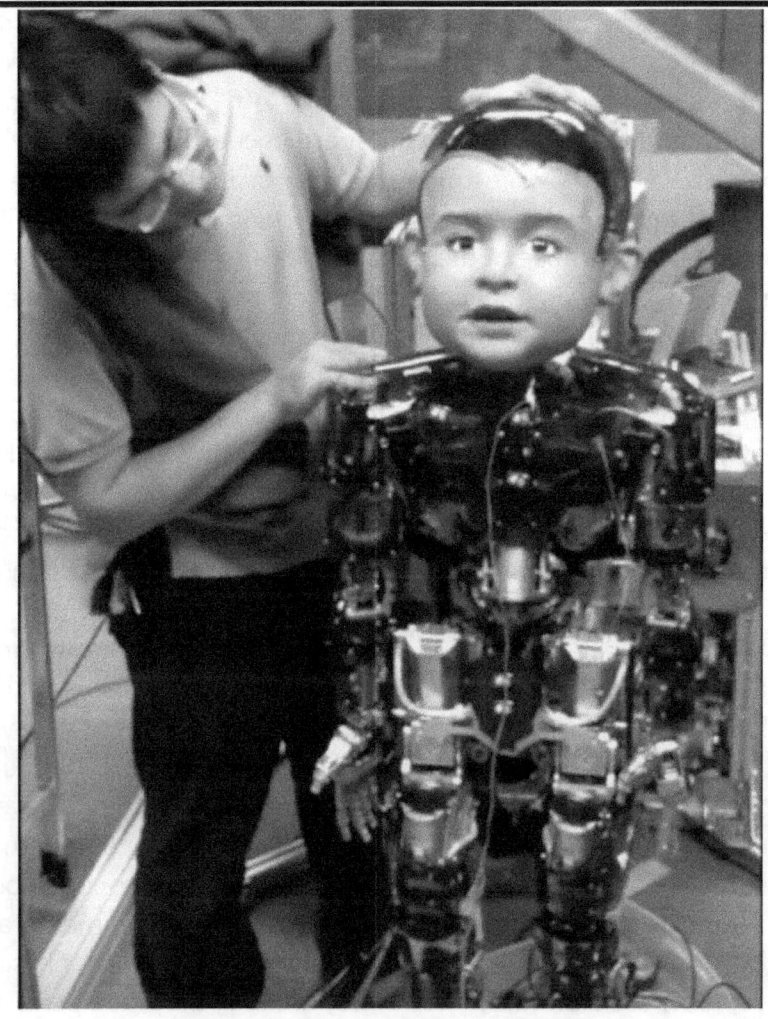

Figure 2: Diego-san Robot (Hanson, 2013)

The graph in Figure 1 attempts to capture the relationship between our comfortableness with a character or robot and its similarity to a human. The curve indicates that more human-like a cartoon character, puppet, or robot is, the more comfortable a viewer is with the character (graph going up and to the right). However, as the character approaches being a perfect human replica, the graph dips. That dip is point where the character's appearance is very human-like but its actions or cognitive abilities do not match how a human moves and reacts in different situations. There is an aspect that is just "off," such as inaccurate eye movements, lip movements that are uncoordinated with what is being said, or hand gestures that do not match what is being said. That imperfection creates an overwhelmingly creepy feeling in an observer because of the mismatch in appearance and action. The dip or valley in the curve is called the "Uncanny Valley," although a better translation from Japanese to English would be the "Valley of Eeriness."

The Uncanny Valley applies to real-world robots. Humanoid robots, such as Pepper, look mechanical and thus signal that they are robots. When humans look at them, they do not seem eerie when they act mechanically because we expect them to act mechanically to match their looks. We do not expect them to have a full, human-like intelligence, so if they respond to us in an odd way, we just expect it is because they are robots. A more realistic, human-looking robot is generally creepy. Hiroshi Ishiguro builds robots that are duplicate images of people. They are given the name "Geminoid" after the mythical Gemini twins. Although they highly resemble humans in appearance, they do not have a correspondingly high human likeness in how they move or in their verbal and non-verbal communication. A Geminoid robot's vocal replies have a bit of delay, their head and eye movements are slightly out of synch with a conversation, and their physical movements are a bit jerky. While they are still, they look perfectly human but, when they move, the robotic movements make the robots appear disturbing.

Designers can exploit the lessons of the Uncanny Valley to make robots that are less hair-raising and instead more interesting than they might be. David Hanson's humanoid robot Sophia often is displayed with the electronics in the back of its head visible. This reduces the creepiness of the mismatch between a very accurate human face and body and slow, stilted verbal and non-verbal communication interactions seen with the Geminoids. Being able to see the wires and circuitry remind viewers that Sophia is a very sophisticated robot and thus people are more impressed with its approximation of a human. Another example of avoiding the Uncanny Valley to make a more interesting, but less life-like robot, is the Sony Aibo dogs. When the Sony Aibo dogs were first being developed in the lab, researchers covered the metallic bodies with fuzzy cloth to make them look more like dogs. As a result, an Aibo dog looked like a wind-up toy dog and people were not impressed. When they removed the cloth costume, the robot's appearance clearly indicated that it was a complex mechanism.

8.2 As You Read "Evidence"

As you read the story, watch for any signs that Byerley is a robot, including his verbal and non-verbal communication. Think about ways robots and humans differ that would help determine whether or not Byerley is a robot without invading his privacy. Note the differences in Calvin's responses in the meeting with Quinn and Lanning, her response to the reporter after Byerley's public speech, and her responses in the private meeting with Byerley. Also think about Calvin's past interactions with robots, including Herbie and Brain, and decide if her interactions with Byerley more closely resemble her interactions with other humans or her interactions with other robots.

8.3 After You Read "Evidence"

Byerley cleverly passes the test that he is a human, though we know he is not. There is really nothing in his actions that give him away, so he is literally on the right side of the Uncanny Valley. Byerley would be a true artificial general intelligence that is often discussed in the press and books warning against robot uprisings.

Asimov's stories usually deal with humanoid robots but the Uncanny Valley applies to robots that do not resemble humans, such as Boston Dynamic's Spot or ETH Zurich's ANYmal four legged robots. These robots look like animals. Their torsos resemble the body of a quadruped but their heads do not look like anything in nature, and may consistent only of a sensor or a manipulator arm or it may even be entirely absent. Their movements are also more mechanical. Most people find Spot and ANYmal technologically fascinating but creepy.

Robot designers are understandably obsessed with building functional robots so they prioritize physical robot morphologies that make a robot more useful at the expense of falling into the Uncanny Valley. If the robot is more cost-effective than another robot (or person), then who cares what it looks like? After all "form follows function" is a fundamental design principle.

But if the robot is going to be working around people, how those people react to the design of the robot is an important function. Think about consumer products like smart phones, kitchen appliances, and cars that are designed to be attractive and eye-catching as well as functional and cost-effective. Those products' morphologies have to combine both form and function in order to be commercially successful.

"Evidence" adds another HRI principle to the list:

- **P11**: *Designers must be aware of the Uncanny Valley and match a robot's appearance and other communication cues to its actual level of intelligence.*

Principle 11 builds on three previous principles. Following Principle 2: Communication between humans and robots can take many forms beyond a graphical user interface, including verbal and non-verbal communication, it acknowledges that non-verbal forms of communication are important. The Uncanny Valley is, to some degree, the result of Principle 6: Humans attribute intent to robots, regardless of whether the robot is acting with intent. Ideally, designers would avoid the Uncanny Valley because they produced robots following Principle 10: A human-robot interaction scheme is based on understanding the interactions between work and environment, agents and relationships, and appropriate communication mechanisms. Yet there are many unintentionally frightful robots, so the Uncanny Valley merits its own principle.

The next and final story in the *I, Robot* collection, "The Evitable Conflict," also features Stephen Byerley. This time, the story will pit him against artificial intelligence software systems called Machines and explore the concepts of autonomy, initiative, and trust.

8.4 Questions:

1. Describe the Uncanny Valley in your own words.

2. Find videos of the following robots: Roomba, Sony Aibo, Packbot, Honda Asimo, ANYmal, and mini-Spot. Place them where you think they would fall on the Uncanny Valley curve.

3. Find videos of the robot Sophia with and without her wiring exposed. What is your emotional reaction to each version—is there a difference? Where do you think each version falls on Mori's graph?

4. Should an artificial general intelligence like Byerley ever be given the rights of a citizen or an elected official? How do you feel about Saudi Arabia making Sophia a "citizen?" Knowing how far AI has to go to create robots like Byerley in the story, should robots like Sophia that represent where

we are right now be allowed to vote or run a town?

5. Watch the explanation of the Uncanny Valley in the "Succession" episode of the TV series 30 Rock where one of the writers tries to explain to Tracy Morgan why his idea for an animated video game will not work. Is this an accurate explanation?

6. Find pictures of the Cybermen and Daleks, two species of villains from the British TV series Dr. Who. Read the entry at http://tvtropes.org/pmwiki/pmwiki.php/Main/UncannyValley, which argues that humanoid Cybermen are scarier than the more mechanical looking Daleks. Do you agree?

References and Suggested Further Reading

Hanson, D. (2013, January 6). Diego installed [YouTube post]. Retrieved from https://www.youtube.com/watch?v=knRyDcnUc4U

Hsu, J. (2012, June 12). Robotics' uncanny valley gets new translation [Online article]. Retrieved from https://www.livescience.com/20909-robotics-uncanny-valley-translation.html

Mori, M., MacDorman, K. F., & Kageki, N. (2012). The uncanny valley [from the field]. *Robotics & Automation Magazine, IEEE,* 19, 98–100. Doi: 10.1109/MRA.2012.2192811

TV Tropes. (2013) Uncanny valley [Web article]. Retrieved from http://tvtropes.org/pmwiki/pmwiki.php/Main/UncannyValley

Chapter 9
Autonomy, Initiative, and Trust: "The Evitable Conflict"

The very title of "The Evitable Conflict" is a hint of Asimov's attitude towards fears of a robot uprising. A robot uprising is not an inevitable or unavoidable conflict that will cause human suffering in the future, but instead it is cycles of war, famine, natural disasters, disease, and other events such as government collapse, that have created suffering throughout history. Robots, or more precisely the machine intelligence that enables the robots, could help mitigate these events by acting as super-intelligent public servants. They could optimize the planning, preparation, and resiliency needed to avoid the consequences of external events beyond our direct control and ability to predict. In this story, the Machines, which are four regional autonomous supercomputers, do just that: They make currently inevitable conflicts evitable or avoidable. The dilemma is that the Machines may, or may not, have exercised a bit of initiative to play dirty so as to better direct the economy and critical infrastructure. So, should we trust the Machines the way we do nameless, faceless—but human—public servants? "The Evitable Conflict" provides a foundation to discuss these questions of autonomy, initiative, and especially trust.

9.1 Before You Read "The Evitable Conflict"

The Machines are basically robots without bodies, a single human master to whom they report, or any pretense of a social interface. This may cause you to wonder if the Machines can be considered robots and if there is any relevance for human-robot interaction. The definition of a robot is debatable, and robotics researchers typically reserve the term for "physically instantiated agents" which means the intelligent system physically interacts with the world. A spaceship like HAL in the movie 2001: A Space Odyssey is sometimes considered a robot because it physically control the ship but sometimes as a software agent. From an artificial intelligence perspective, there is very little difference between a physical agent and a software agent: they both require a wide range of artificial intelligence capabilities and the software is organized in a similar fashion.

As to relevance for HRI, the Machines in the story expand on many ideas already presented in previous stories. The Machines are autonomous, which was defined in the introduction. They have a high level of initiative, a term we discussed in "Catch That Rabbit." They work on ethically sensitive applications because of the impact of the policy decisions. They also exhibit full moral agency, as described in "Liar!" They have no need for bodies so they avoid the Uncanny Valley effect that we examined earlier in "Evidence." At the point in the future where this story takes place, people trust the Machines with important decisions, and this trust is different from the interpersonal trust or manipulation of trust that we saw in "Robbie." This chapter will focus on how humans trust systems with high autonomy and initiative, which will, in turn, help us understand and trust regular systems such as bomb squad robots, robot vacuum cleaners, and delivery assistants.

9.2 As You Read "The Evitable Conflict"

As you read the story, watch for signs of bounded rationality on the part of the Machines. Are there limits on how well they can compute solutions for mankind's societal problems due to external forces? For example, do they have enough information, computing power, and time to make effective decisions? What level of initiative do the Machines exercise in order to act on their projections of how to mitigate the negative effects of external events? Also be on the alert for indications as to whether regular people without a robotics background trust the Machines, and how these people formed their opinions of the Machines. At the same time, consider whether robot experts such as Calvin and Byerley trust the Machines and whether their reasons for trusting or not trusting them would be different than the reasons formed by regular people without their expert knowledge.

9.3 After You Read "The Evitable Conflict"

Do you trust AI and robots more or less after reading the story? The Machines were designed to maintain economic stability. When small patches of unemployment or other abnormalities appeared, Byerley assumed that a bug in the programming created the problem. Of course, the story indicates that lesser machine intelligences constructed the Machines' AI, similar to the way the Brain constructed the spaceship in "Escape," so that no one or thing, really knew how it worked. In this case, when Byerley goes to the Machines to question them about the reason for the anomalies, they refuse to answer him. After his investigation, he is highly concerned about the minor fluctuations. Based on Byerley's findings, Calvin explains the problem is the high level of initiative given to the Machines, despite their bounded rationality. Each Machine saw what to them was an optimal solution to a problem of societal importance that would not violate the Three Laws.

Why would anyone trust a black box like a Machine? Well, people are black boxes, too, and we trust them. Stephen Covey, the business management guru, wrote a very influential book on trust, called The Speed of Trust that describes the characteristics that make a person trustworthy. Richard Barrett took these characteristics and created the Trust Matrix graphic shown in Figure 1. The Trust Matrix captures the idea that our trust in a co-worker is influenced by both the person's character and their competence. Good character is about how we trust partnering or teaming with a person, competence is about how we trust the individual's ability to do the job. Both are important.

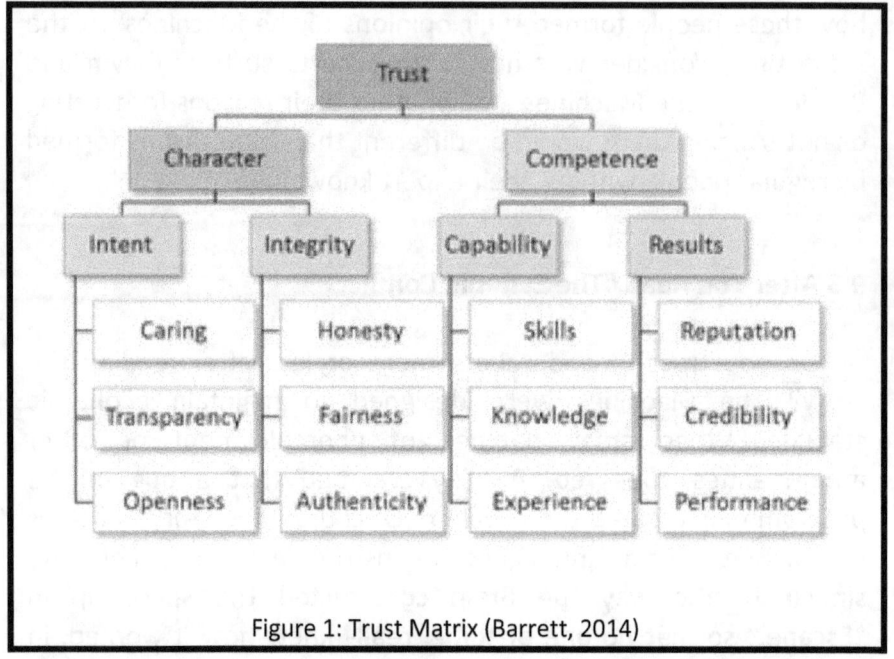

Figure 1: Trust Matrix (Barrett, 2014)

As shown in the left branch of the diagram in Figure 1, a person's character is a function of their intent to work with others and their integrity. A person with good character is honest, fair, and authentic, but as a manager, we also expect them to be caring towards subordinates, transparent in decision making, and open to new ideas.

Of course, good character is not sufficient to get a job done; we also trust a person to complete a task based on their competence (right branch of Figure 1). A hiring manager decides a person is competent based on their past history or resume that tells their skills, knowledge, and experience. We also can tell a person is competent by the results of their previous work: recommendations from others, examining how close past work is to the work they will be doing for us, and what their performance track record looks like. Although someone is hired based on all these things, real, personal trust does not begin until we witness the results for ourself.

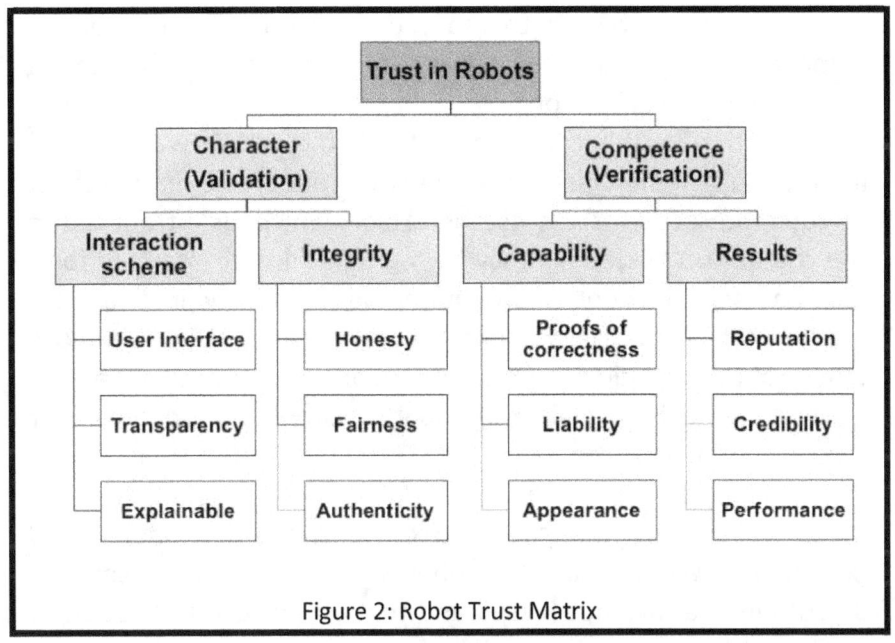

Figure 2: Robot Trust Matrix

Does this translate to our trust in robotics? Yes, though it is not a perfect match. Figure 2 shows the Robot Trust Matrix. Like the Business Trust Matrix, the Robot Trust Matrix supposes that our trust in robots is based on their character and competence. But the use of the term "character" sounds like only a robot that is a full moral agent can be trusted. Instead, it is more useful to think of character and competence in terms of software engineering.

A major software engineering principle is verification and validation. A successful program not only must correctly meet the explicit contractual requirements (verification) but also meet the real intent of the application (validation). We have all downloaded smartphone apps that were without bugs or defects, but were too clunky or slow or just created an awful user experience. We rejected the apps after these poor interactions. This leads to a general rule for trust in software: as programmers, we obviously are required to ensure the program works correctly, but as programming professionals, we want to make sure the program really does a good job at its intended use. But understanding what people want a program to really do is hard. It should be no surprise that, historically, computer science is better at verification than validation.

Validation loosely corresponds to character. We want a good interaction scheme when working with a robot. One major aspect of good interaction is the user interface, which, as introduced in the chapter on "Escape!", may be a graphical user interface for a field and service robot or a multi-modal naturalistic interface for robots working in social settings. We want transparency so we can tell if robot execution is proceeding correctly. We also want to easily be able to explain the robot's actions and to debug or reconfigure it with the same ease. In addition to having a good interaction scheme, we want and expect the robot to have integrity. Following the trust matrix, this means the robot should be honest. We expect the robot to treat us fairly through algorithms that are not biased and that do not use perfection as an ideal. For example, we do not expect a robot to keep reminding us of our past mistakes nor do we expect it to make fun of us when we make new mistakes. We also expect robots to be authentic and not manipulative. For example, we want a robot that truly listens to us and responds appropriately using non-verbal communication to make interactions socially comfortable. No one wants to interact with a robot that rises up as large as it can and flashes red eyes at you every time you try to give it a command or interact with it.

Verification confirms a robot's competence. Whereas a human has a resume that captures internal skills, knowledge, and experience, a robot has other indicators. In Asimov's stories, the robots always have mathematical proofs of correct operation for an application even though the application might not be correctly specified and leads to unexpected robot behaviors! In the stories, the robots come with some sort of warranty or legal liability, and the Three Laws serve as consumer protection standards. In real life, the liability and consumer protection laws for robots are unclear.

Verification is also about observing the results of a robot. Consider the components that make up results in the human trust matrix: reputation, credibility, and performance. We believe a robot works correctly based on its reputation and its credibility. If others have used the robot successfully, we believe it will work for us as well. Similarly, if someone previously used the robot for our specific application, then it is credible that it will work for us. The component of results that is most influential on whether or not we will trust a robot is observing and verifying the robot's performance for ourself. That is also the result that is the most obvious for us to check: Does the robot work correctly during a representative set of tests? Does it continue to work as well outside of the laboratory?

A third indicator, appearance, may seem surprising or out of place. As discussed in the chapters on "Reason" and "Evidence," a robot's morphology influences whether or not a person thinks the robot is capable of the job. For example, during the 9/11 World Trade Center disaster, I personally watched responders decline to use a particular model of robot because it looked too experimental. Instead, they chose a less capable robot model that looked sturdier.

The Machines in the story created a problem of trust. They were competent by every measure, but their character was suspect, at least to robot experts. Their actions and reasoning were not readily transparent or explainable. Thus, there was no way to know if their actions were correct. Because their intelligence was beyond that of humans and their actions were beyond a human way of thinking, it was impossible for human troubleshooting. Oddly enough, neither Byerley or Calvin doubted the Machines' integrity: Whatever the Machines were doing, they were doing it for what they truly thought of as the right reasons. It was never assumed the Machines had altered their programming or somehow erased or superseded the Three Laws. That brought the problem of trust back to trusting in their competence—either the Machines were doing a good job or what they were doing was still better than what humans could accomplish. In either case, their policies and decisions were bounded by the Three Laws.

"The Evitable Conflict" adds this final HRI principle to the list:

- **P12**: *Trust in robots is based competence and character; which loosely corresponds to verification and validation in software engineering.*

"The Evitable Conflict" is both a positive and ambiguous last word on robotics. The Machines are ultimately making everyone's life demonstrably better while not interfering with individual freedoms. And yet, humans have lost some control over their lives by delegating the boring, hard to comprehend decisions of managing society's infrastructure to the Machines. When we see this near utopian society created by the Machines, we must ask, "Is any lingering discomfort in delegating tasks to a robot a sign that we do not trust them? Or is it that we have do not trust ourselves to create good robots?" It is hoped that the answers to both questions would be "no" because designers have embodied the principles of good human-robot interaction into their creations.

9.4 Questions:

1. The Machines made the inevitable conflict between humans and their environment (as well as interhuman conflicts) evitable, but is there an inevitable conflict between AGI and humans? For instance, what if an AGI handled food production and distribution differently and forced everyone to become vegans? Could conflicts be avoided? Would the Three Laws be enough?

2. Would you trust a medical robot like a DaVinci surgical robot to perform surgery on you? Would you trust a legged robot like the Boston Dynamic Spot robot to patrol the area in which you are working for safety concerns? Why or why

not? If Spot robots were as common as Da Vinci robots would your opinions change?

3. Did the Machines act with integrity considering that they damaged the reputations of people who were in their way? Were the Machines' actions to neutralize dissent a function of their competence or a failure of their ethical integrity?

4. The Machines are full moral agents who lied. Is there any difference between them and Herbie in the story "Liar!"?

5. In "The Evitable Conflict," is humanity's complete trust in the Machines warranted?

6. The Machines' actions could be interpreted as a very subtle robot uprising. On the other hand, the Machines are similar to a group like the Federal Reserve or the International Monetary Fund who manipulates the U.S. economy and other economies to keep them on an even keel. In light of this information, do you think "The Evitable Conflict" is a story about a robot uprising? Why or why not?

References and Suggested Further Reading

Barrett, R. (2014, April 11) Building trust in your team: The trust matrix [Web blog post]. Retrieved from https://richardbarrettblogdotnet.wordpress.com/2014/04/11/building-trust-in-your-team-the-trust-matrix/

Covey, S. M. R. (2006). The speed of trust: The one thing that changes everything. New York: Free Press.

Chapter 10
Conclusion

The *I, Robot* stories are remarkable given their age. The stories continue to entertain modern audiences, and the stories hit upon some pervasive issues concerning how we as humans want to interact, and how we actually interact, with robots. The stories provided us with imaginary case studies from which we developed principles in HRI, which are collected below into a single list for convenience. Of course, the stories do not cover the entirety of human-robot interaction. They missed three important aspects of HRI, also added below. However, this book was not meant to be an exhaustive text on the topic. Hopefully, the information has whet your appetite to learn more about HRI and artificial intelligence. To further your knowledge, the chapter ends with suggestions for additional reading.

10.1 Filling in the Gaps

I, Robot stories typically featured robots that are far beyond the capabilities of real-world robots. These robots exhibited artificial general intelligence (AGI) with naturalistic communication modes as they served as taskable agents for a variety of applications. The current state of robotics is much different and these gaps influences research directions in HRI. Similarly, while the stories discussed how the robots were designed, how users interacted with them, and especially how they were debugged, nothing was ever said or implied about the methods for designing and testing human-robot interaction.

	Featured in *I, Robot*	Overlooked by *I, Robot*
MORPHOLOGY	humanoid	non-anthropomorphic
AUTONOMY	full	shared
INTELLIGENCE	artificial general intelligence (AGI)	narrow intelligence
COMMUNICATION	naturalistic	physical, graphical, BMI
DESIGN LIFECYCLE	operations	development

Figure 1: What *I, Robot* Missed About HRI

In real life, robots are built for specific applications, not for general purpose use. The majority of robot applications fall into two categories: those for a highly specialized field, such as manufacturing and mining robots, and those for service applications, such as warehouse robots, vacuum cleaners, drones for mapping farms or areas impacted by a disaster, or self-driving cars. Only a few modern robots are designed with the intent of producing AGI, such as Sophia, and those are research prototypes. Social robots, such as entertainment robots or robot tutors, are emerging in the commercial market but do not have significant market share. Interactions with specialized and service robots typically rely on physical user interfaces (e.g., buttons, joysticks, graphical displays) and rarely limited natural language comprehension. Thus, a big gap in human-robot interaction is improving interactions for field and service applications.

Unlike *I, Robot*, modern robots are rarely completely taskable agents but often share control of tasks with a human which poses another set of challenges for HRI. The best example of this is autonomous driving cars. The cars are not fully autonomous as the person is expected to take over from time to time or to be supervising the robot's driving in case that something goes wrong (remember the exception handling phase introduced in the chapter on "Runaround"). Human factors researchers have known from studying crashes in airplanes with autopilots in their early years of use that people who are not actively engaged in a task do not pay attention or simply cannot respond in time to these problems. This is called the human out-of-the-loop control problem and is something that needs to be addressed as we work toward creating reliable human-robot interaction.

In all of the *I, Robot* stories, not only do the robots perform their tasks correctly most of the time without supervision but also we see troubleshooters having an excessive amount of time to correct problems that arise. Research in human-robot interaction is addressing the human out-of-the-loop control problem and how to help people react appropriately to problems with robots.

The *I, Robot* stories also failed to reveal the unglamorous design activity of conducting human-robot studies to understand human-robot interaction. This activity requires researchers to develop metrics from those studies to quantify if the HRI scheme is really working, which has proven to be a very hard problem. If you look at the published papers on human-robot interaction, most of them study and collect data with a large number of people using a robot under highly controlled circumstance in order predict quantitatively how people will interact with robots. Setting up and running this kind of study is often outside the training that an engineer or computer scientist gets. Even if a study on HRI is conducted, researchers often use volunteer psychology students in order to get a large enough number of people for the study to have statistical significance. The students are not guaranteed to react the same way a senior citizen would or a law enforcement agent. Furthermore, the highly controlled scenarios rarely incorporate all the ways a robot can fail or do something unexpected so the studies may miss important aspects of the interaction. Fortunately, human-robot interaction research methodologies are under development and will continue to improve.

10.2 Principles

The nine stories from *I, Robot* illustrate 12 basic principles of how robots work and how people expect these robots to act during human-robot interactions:

- **P1:** *What is easy for a human is hard for a robot, and what is easy for a robot is hard for a human.*

- **P2:** *Communication between humans and robots can take many forms beyond a graphical user interface, including verbal and non-verbal communication.*

- **P3:** *A complete human-robot interaction scheme enables humans to appropriately interact with a robot in each phase of its task cycle (initiation, execution, termination, exception handling), even if the robot is a taskable agent.*

- **P4:** *Ignoring the design of a human-robot interaction scheme to facilitate interactions does not mean that there will be no interaction effects, only that accidental interaction effects will likely be unsatisfactory.*

- **P5:** *Adding more intelligence to a robot does not eliminate the need to design the robot to facilitate positive human-robot interactions.*

- **P6:** *Humans attribute intent to robots regardless of whether or not the robot is truly acting with intent.*

- **P7:** *Every robot, even those not used for ethically-sensitive functions, has ethical aspects for which the designer is responsible.*

- **P8:** *It is difficult to create or debug an effective HRI scheme without understanding how robots work.*

- **P9:** *A user interface can be either for end-user execution, for a developer's diagnostics, or for explication to the public.*

- **P10:** *A human-robot interaction scheme is based on understanding the interactions between work and environment, agents and relationships, and appropriate communication mechanisms.*

- **P11:** *Designers must be aware of the Uncanny Valley and match a robot's appearance and other communication cues to its actual level of intelligence.*

- **P12***: Trust in robots is based competence and character; which loosely corresponds to verification and validation in software engineering.*

The above list is not guaranteed to be exhaustive. This should be expected since the stories focused on robots with artificial general intelligence. Even though it may be incomplete, this list should remind engineers, psychologists, and communications experts of the complexity of intelligence and the complexity of interactions between intelligent agents. The list and stories themselves remind us that that no single discipline has the key to predicting and designing a desirable human-robot interaction schemes. It is through cross-disciplinary efforts that robots will evince more favorable interactions.

11.3 Further Reading

Further reading in human-robot interaction is stymied by the lack of an accepted textbook. HRI as a field of study has only been in existence since 2002. It usually takes at least 20 years for researchers to agree on the breadth or scope of a field and what is important about it. Numerous journals have emerged that publish studies on human-robot interaction, but they are very technical. Conference publications are often easier to read. There are two particularly noteworthy conferences: the *ACM/IEEE International Conference on Human Robot Interaction* (nicknamed "HRI," which can be confusing) and the *IEEE International Symposium on Robotics and Human Interactive Communication* (nicknamed "RO-MAN"). HRI tends to favor laboratory studies and social issues, while RO-MAN attracts more studies based on users in the field conducting applications.

In addition to the suggested reading found throughout this book, if you want to learn more about artificial intelligence in general, my favorite book is *AI Rebooted: Building Artificial Intelligence We Can Trust* by Gary Marcus and Ernest Davis, (2019). *AI Rebooted* is nominally about how deep learning is not "the answer" to creating artificial intelligence. It also explains how deep learning will not eliminate all the other areas of AI research that have existed since the 1960s. Along the way, they give an excellent overview of AI, the different concepts involved in creating it, why these concepts are important, and a general history of the field. Also, the authors use humor that borders on snarkiness to keep the material from getting too dry.

If you want to learn more about artificial intelligence specifically for robotics, there are two light options at the level of a PBS or Discover Channel special. My book *Robotics Through Science Fiction: Artificial Intelligence Explained Through Six Classic Robot Short Stories*, (2018) is arranged much like this book with science fiction stories to illustrate the material. However, that book covers the details of how robots are programmed in more depth. Another wonderful tongue-in-cheek tome is *How to Survive a Robot Uprising* by Daniel H. Wilson (2005). Wilson is a science fiction writer who has a Ph.D. in robotics from Carnegie Mellon University. Even though it was written in 2005, most of the concepts still apply. If you are looking for a heavier textbook, I am partial to my own: *Introduction to AI Robotics* (2019). It is a survey for college seniors or first-year graduate engineering students. Although it minimizes the actual mathematics and programming knowledge you need to have, it is not meant for light reading.

Regardless of what you read next, the hope is that you will keep learning about this amazing field!

ABOUT THE AUTHOR

Dr. Robin R. Murphy is the Raytheon Professor of Computer Science and Engineering at Texas A&M. She is one of the founders of the field of human-robot interaction, an award winning textbook author, and a TED talk speaker. The Best Master's Degrees declared her one of the 30 Most Innovative Women Professors Alive Today. She is considered one of the most influential women in technology. Her interest in human-robot interaction resulted from her field work using robots for disaster response. While responding to disasters such as the 9/11 World Trade Center attacks, Hurricane Harvey, and the Fukushima Daiichi nuclear accident, she documented that although the robots worked physically well, there was an unusually high rate of human error, frustration, and fatigue.

Murphy frequently appears on CNN, NBC, NPR, Popular Science, NY Times, and in the popular press. As an Innovative Teaching Faculty Fellow at Texas A&M, she pursues more engaging forms of education, particularly through the use of science fiction to enable students to better visualize abstract concepts in artificial intelligence, understand how the algorithms actually work, discover what would be the impact on systems design, and explore the ethics of artificial intelligence. This resulted in her Robotics Through Science Fiction blog, her book 2018 book *Robotics Through Science Fiction: Artificial Intelligence Explained Through Six Classic Robot Short Stories,* and her ongoing column on science fiction and science fact for *Science Robotics*.

www.ingramcontent.com/pod-product-compliance
Lightning Source LLC
Chambersburg PA
CBHW070654220526
45466CB00001B/439